THE SCIENTIFIC WORLD OF COPERNICUS

NICOLAI CO-
PERNICI TORINENSIS
DE REVOLVTIONIBVS ORBI-
um cœleſtium, Libri VI.

Habes in hoc opere iam recens nato, & ædito, ſtudioſe lector, Motus ſtellarum, tam fixarum, quàm erraticarum, cum ex ueteribus, tum etiam ex recentibus obſeruationibus reſtitutos: & nouis inſuper ac admirabilibus hypotheſibus ornatos. Habes etiam Tabulas expeditiſsimas, ex quibus eoſdem ad quoduis tempus quàm facillime calculare poteris. Igitur eme, lege, fruere.

Ἀγεωμέτρητος οὐδεὶς εἰσίτω.

Norimbergæ apud Ioh. Petreium,
Anno M. D. XLIII.

THE SCIENTIFIC WORLD OF COPERNICUS

On the Occasion of the 500th Anniversary of his Birth
1473–1973

Edited by

BARBARA BIEŃKOWSKA
University of Warsaw, Poland

Foreword by

ZDENĚK KOPAL
Victoria University of Manchester

D. REIDEL PUBLISHING COMPANY

DORDRECHT-HOLLAND / BOSTON-U.S.A.

Published by arrangement with Agencja Autorska, Warszawa
Translated from the Polish by Christina Cenkalska

Library of Congress Catalog Card Number 73–85712

ISBN 9 0277 0353 1

Published by D. Reidel Publishing Company,
P.O. Box 17, Dordrecht, Holland

Sold and distributed in the U.S.A., Canada, and Mexico
by D. Reidel Publishing Company, Inc.
306 Dartmouth Street, Boston,
Mass. 02116, U.S.A.

Printed in The Netherlands by D. Reidel, Dordrecht

TABLE OF CONTENTS

FOREWORD

On February 19, 1973, five centuries have elapsed since the birth of Nicolaus Copernicus – the greatest astronomer of the Renaissance period – who rediscovered for us the heliocentric model of the solar system, and documented it by his life's work in such a manner as to make its concept a permanent property of mankind.

The life of Copernicus, extending from 19 February 1473 to his death on 24 May 1543, was not too rich in adventures or biographical facts. Born in Toruń from a family of Polish burghers, he received his first university training in Cracow between 1491–1494. From Cracow he proceeded to Italy to spend the years between 1496–1503 at the Universities of Bologna, Padua and Ferrara – with occasional visits to Rome – in preparation for an ecclesiastical career. When Bishop Watzenrode – his patron and maternal uncle – could no longer extend his leave, Copernicus returned to Poland in 1503 to enter the service of the church establishment, which soon led to a canonry at the Frombork (Frauenburg) Cathedral in Warmia. And there – in the northern mists not far from the Baltic shores – in a land so different in climate from the sunny Italy of his youth – he was destined to spend the rest of his life.

Copernicus' interest in astronomy was probably awakened already at Cracow by his first outstanding teacher, Wojciech of Brudzewo; though it was not till during his Italian years that he conceived the design which was to blossom out into the finest flower which astronomy contributed to the human culture at the time of the Renaissance. Relatively little is known of Copernicus' life during his Italian years; and after five centuries his trail is difficult to follow from extant records with any assurance. From the early years of his life Copernicus had an inclination to anonymity which grew in him with the years. Only now and then – when contemporary events happened to place him temporarily in a position of prominence – do we recognize his features more clearly; but soon thereafter he lapses back into the shadowy existence which was his second nature. His location, at Frombork, in "the most remote corner of the Earth" – as Copernicus described it – precluded more intimate personal contact with many other scholars of his time; and his ecclesiastical rank (though Copernicus was a canon – and between 1523–1525 chancellor of the Frauenburg chapter – he was never ordained priest) deprived him of a family which he could call his own.

There is, however, no doubt that during his 12 years of University life – both in Poland and abroad – he learned in his youth all that ancient sources had to offer – and these were still the principal sources of human knowledge at that time. It is, therefore, more than probable that, on his return to Poland from Italy in 1503, he brought back with him at least the germ of the idea whose execution was to occupy him for the rest of his life, and to immortalize his name in the annals of our science: namely, a synthesis of the heliocentric model of the solar system of Aristarchos with the geometrical apparatus of Ptolemaic astronomy.

The idea of the heliocentric model of the solar system was not born at the time of the Renaissance, but goes back much further in the cultural history of mankind. In fact we owe it – like so much else – to the Greeks of the Hellenistic times – times when internal disunity lost them their political power, but when they became the masters of the intellectual world by the sheer weight of their genius. As far as we know, the idea that the Sun – rather than the Earth – is at the centre of our planetary system, and that the Earth revolves around it, was propounded first by Aristarchos of Samos – an outstanding philosopher of the third century B.C. – whom astronomers can rightfully adopt as one of their early patron-saints. Of his life much less is known to us than we know about Copernicus. His date or place of birth is unknown – probably around 310 B.C. – the only fixed date of his lifetime being the year 281 B.C., when he observed (according to Hipparchos) the time of the summer solstice. We know that he was berated for the boldness of his views by the stoic philosopher Kleanthes some time after 264 B.C. We have, however, no idea where Aristarchos lived – whether in Athens or Alexandria – nor where he died; but most part of his life was probably spent in the first half of the 3rd century B.C.

What is even worse – Aristarchos' views on the structure of the solar systems may have remained unwritten or what he wrote was lost, so that we know about it only from the references to his views made by his contemporaries. The most famous among these is the testimony recorded by the great mathematician Archimedes (287–212 B.C.) in his *Psammites* (Sand Reckoner) – in words which cannot be read without emotion even after the lapse of 22 centuries:

> You (King Gelon II, tyrant of Syracuse, who died before 216 B.C.) are aware that universe (κοσμος) is the name given by most astronomers to the sphere whose centre is the centre of the Earth, and whose radius is equal to the distance between the centre of the Sun and the centre of the Earth. This is the common account as you have heard from astronomers. But Aristarchos of Samos brought out a book consisting of some hypotheses wherein it appears, as a consequence of assumptions made, that the (real) universe is many times greater than the one just mentioned. His hypotheses are that fixed stars and the Sun remain unmoved, that the Earth revolves about the Sun in the circumference of a circle, the Sun lying in the middle of the orbit, and that the sphere of the fixed stars, situated about the same centre as the Sun, is so great that the circle in which he supposes the Earth to revolve bears such a proportion to the distance of the fixed stars as the centre of the sphere bears to its surface ....

To express the meaning of this passage in plainer words, Aristarchos attempted to explain the observed astronomical phenomena by postulating the daily rotation of the Earth about its own axis, as well as the yearly revolution of the Earth around the Sun. A hypothesis of the terrestrial daily rotation was, to be sure, advanced previously by Heracleides; but in postulating the second (annual) motion Aristarchos does not seem to have had any predecessors. Moreover, the existence of such motion would explain in one stroke all retrograde loops exhibited by the outer planets (Mars, Jupiter and Saturn) as a reflex of our own yearly motion around the Sun.

This being said, the reader may ask why the heliocentric planetary system was not adopted readily in antiquity, and had to wait another 17 centuries for Copernicus.

The reason is the fact that the simple model envisaged by Aristarchos failed to represent the observations ("save the phenomena", as the Greeks were wont to say), within the limits of errors of even naked-eye observations; and this could not have been otherwise because of another shortcoming which took many centuries to shake off: namely, the pre-conception that the motions of all celestial bodies must necessarily be circular and uniform.

We do not know who planted this particular false seed in the human mind – probably the Pythagoreans, misled by esthetic rather than physical reasons. But its consequences for the progress of science were singularly disastrous. For we know, of course, and have known since the time of Kepler, that planetary motions in the sky are neither uniform nor circular; and that any attempt to represent the apparent motions of the planets in the sky by pure circles would fail to establish a real agreement between theory and observation regardless of the position of the Sun in the system. This is especially true of our nearest celestial neighbour Mars, which happens to follow a markedly eccentric orbit ($e = 0.093$) – and for which, as a result, the differences between the observations and any system – be it geocentric or heliocentric – based on the assumption of uniform circular motion would amount to whole degrees.

Discrepancies of this order of magnitude must have troubled Aristarchos, and been intolerable to Hipparchos living a century later. Therefore, the heliocentric model of the solar system, introduced by Aristarchos in the 3rd century B.C., failed no doubt to gain acceptance because a mere replacement of the Earth by the Sun at its centre – simplify as it did some geometrical aspects of the problem – did nothing to remove the principal obstacle to the practical usefulness of such models: and this was the mistaken concept that all celestial motions must be circular and uniform. There is no indication known to us that Aristarchos would have contemplated any departure from the Pythagorean tradition in this respect; nor did anyone else until the time of Kepler.

In the meantime, the only way which suggested itself to the ancient astronomers for 'saving the phenomena' and reconciling the observed paths of the planets on the celestial sphere with the Pythagorean preconceptions, was to superpose more than one uniform circular motion on top of another. The commencement of this geometrical merry-go-round can be traced to the homothetic spheres of Eudoxos, followed by the geometrical theory of epicycles laid down by Apollonios in the second half of the 3rd century B.C., which was developed further by Ptolemy into a pragmatic geometrical system which survived in astronomy – with many vicissitudes – until the time of Copernicus.

Was Copernicus aware of the fact that the heliocentric system of the planetary family was already proposed in antiquity by Aristarchos? During his Italian years Copernicus no doubt learned all that contemporary astronomy had to offer; and – if nobody else – Mario Novara in Padua would have made him aware of the various shortcomings of the geocentric system. The 'editio princeps' of Archimedes' *Psammites*, with its crucial testimony, quoted above, appeared in Basle in 1544 – one year after Copernicus died. However, that Copernicus was aware of Aristarchos' work from manuscript sources is attested by his own hand in the manuscript of his book *De Revolutionibus Orbium Celestium.* The respective passage was eventually left out of the printed edition of his work by a stroke of the author's pen; but the manuscript containing it has been preserved to this day.

Copernicus doubtless knew too from his teachers – and, may be, partly from his own observations – that the heliocentric system based upon the Pythagorean preconceptions on the uniformity of motions cannot be made to represent the apparent motions of the planets in the sky even within the limits of accuracy of the observations made with naked eye. This is why, in order to improve agreement, Hipparchos and Ptolemy felt impelled to introduce their complicated systems of epicycles. Copernicus could offer no more adequate mathematical machinery to the same end; but in place of applying it to geocentric orbits of celestial bodies, he set out to graft it on to a heliocentric system. Copernicus was the first one to attempt such a synthesis; and his stature in the history of science rests on this fact. Moreover, the merits of such an accomplishment are scarcely diminished by an admission that Copernicus did not invent the heliocentric model of the solar system, but revived its ancient tradition going back to pre-Ptolemaic days. His main personal contribution was – we repeat – to dress up the heliocentric idea of Aristarchos in the garb of the Ptolemaic geometry. He could scarcely have done otherwise, for kinematically Copernicus was still fully under the spell of the Pythagorean lore; but he was the first to develop such a synthesis into a comprehensive system.

Before we say a few words on the actual accomplishments of the resulting system, let us describe first some of its salient features. Thus – perhaps contrary to many popular views – the geometrical centre of the Copernican model of the solar system did not coincide with the position of the Sun, but rather with the centre of the circular orbit of the Earth; for it was only in this way that he could account for the phenomena arising (as we now know) from the eccentricity of the terrestrial orbit. Secondly, in order to reconcile his geometrical model of the solar system with the astronomical observations (made mostly by others, to which Copernicus added very few of his own) within 10 minutes of arc – this was the goal of Copernicus' efforts – he had to employ actually a larger number of epicycles to this end than were used in the geocentric models of the same period. The latter, in the hands of George Peurbach (1423–1461) required only 40 epicycles to 'save the phenomena' to a comparable accuracy; while the final system of Copernicus of 1543 required not less than 48 of them; and by abolishing the 'equants' from his theory he had to introduce more 'deferents' in their place.

Under these circumstances, it is only to be expected that the overall picture of the Ptolemaic system of the 16th century was really no more complicated geometrically than that of Copernicus. The latter also did not represent any significant advance in our knowledge of the true size of the solar system. Whereas for Ptolemy (who lived in the 2nd century A.D.), the distance to the Sun was estimated to 610 Earth diameters, Copernicus diminished this distance to 571 Earth diameters (the actual value of the 'astronomical unit' being 11 500 Earth diameters). Moreover, while in the geocentric model the ratios of planetary distances are essentially arbitrary, the heliocentric model contains a 'built-in' system for a determination of the relative sizes of planetary orbits in terms of that of the Earth (the annual retrograde motion of each planet being merely a reflex of our own motion around the Sun). In this way, Copernicus constructed for the first time the actual model of our solar system; though he still underestimated its scale by a factor close to 20.

It has, however, been traditional in the science of astronomy to measure the value of a new planetary theory by the correctness with which such a theory can predict the positions of the planets in the sky at future times. The planetary tables constructed by Copernicus on the basis of his heliocentric theory (published by Erasmus Reinhold in Wittenberg in 1551, eight years after Copernicus' death, under the title of *Tabulae Prutenicae*) did constitute a significant improvement over the preceding *Alphonsine Tables* from the 13th century, based on the geocentric system. For three quarters of a century the Copernican tables of planetary motion remained the standard source of information – until they were superseded by the *Rudolphine Tables* of Johannes Kepler in 1627; but, by that time, the entire astronomical scenery had undergone a profound change, and the work of Copernicus became a part of history.

Tradition has it that Copernicus received the first copy of his great work *De Revolutionibus Orbium Celestium* (Nürnberg, 1543) on his deathbed, barely aware of its contents. How was the synthesis of the idea of Aristarchos presented in terms of the Ptolemaic geometry received by subsequent generations – in particular, among the clergy – men concerned with philosophical implications of the system? Its thesis was accepted with lukewarm interest on the part of the educated Catholic clergy, and without demur by Pope Paul III to whom Copernicus' book was dedicated. On the other hand, the Lutheran church objected strongly from the beginning to its thesis on grounds of biblical fundamentalism. Martin Luther himself gave out only a few uncouth growls in this connection ("That fool wants to turn the whole art of astronomy upside down") in his native vernacular; while Melanchthon proved the Earth to be at rest in elegant Latin. The real theological storm – in which both Catholics and Protestants began to outbid each other in their denunciations of the heliocentric system – did not break out till the first 3rd of the 17th century, in the wake of the work of Galileo Galilei and Johannes Kepler.

It was, in particular, Kepler – that David armed with mathematics rather than stones in his sling – who was destined to slay that Goliath of Pythagorean preconceptions, and to liberate the planetary theory from the straightjacket of uniform motion in circular trajectories. The dogma of such motions was, however, still held as sacrosanct by Copernicus. For listen to what he had to say on this subject in his *Commentariolus* of 1574:

> Our ancestors assumed a large number of celestial spheres for a special reason: to explain the apparent motions of the planets by the principle of regularity. For they thought it altogether absurd that a heavenly body should not always move with uniform velocity in a perfect circle....

And, as these Pythagorean tenets failed to 'save the phenomena' to the desired accuracy, Copernicus went on ...

Having become aware of these defects, I often considered whether a more reasonable arrangement of circles could perhaps be found ... in which everything would move uniformly about its proper centre, as the rule of absolute motion requires.

As the repudiation of such views, held without any foundation for so many centuries, was a *condito sine qua non* for a true renaissance of our science, Copernicus can scarcely be regarded as an actual forerunner of men like Galileo or Kepler, who a century later were to call themselves Copernicans, but whose bold departures from tradition would have horrified their master. No, Copernicus was not one of them. He was, instead, the last of the great ancients – a spiritual companion of Aristarchos, Hipparchos or Ptolemaios from whom he was separated by many centuries of time, but with whom he shared their outlook of the heavens more fully than he would have done with Galileo or Kepler on the other side of the 'great divide' after 1610.

By that time, the work of Copernicus belonged already to the history; and in the history of astronomy his place is forever secure.

Manchester, 1973 ZDENĚK KOPAL

THE COUNTRY AND THE WORLD OF COPERNICUS

STANISŁAW HERBST
University of Warsaw

NICOLAUS COPERNICUS, Torunii in Prusia natus, patre Nicolao Copernico, matre vero quae erat germana soror Lucae a Watzelrod Toruniensis, episcopi Varmiensis, praeclare de Republica Polonorum meriti in causa Cruciferorum.*

Copernicus was born in Toruń (Thorn) and quite likely attended school at the neighboring town of Chełmno. Except for five years as student in Cracow and eight in Italy, Copernicus spent the adult years of his life at the bishopric of Lidzbark as canon of the diocese of Warmia (Ermland), at the chapter's administrative center in Olsztyn and, the longest period, at the cathedral of Frombork. It is from here that he traveled to the provincial councils of Prussia and made two trips to the capital of Poland Cracow and perhaps one to Poznań.

His country was the province of ancient Royal Prussia, composed of his native Toruń and Warmia, both components of the Polish state since 1454. In the largest sense, however, one might say that Copernicus was a citizen of the universal republic of scholars, a member of a community that was representative of European culture at the turn of the 16th century.

1. His Country

In the second half of the 15th century, the kingdom of Poland, united by treaty with the Grand Duchy of Lithuania in 1386 and since then ruled by the Jagiellonian dynasty, was the largest country of Europe. Owing to the decline of Mediterranean commerce after the Turks seized Constantinople in 1453, the bulk of European trade was shifted to the Baltic and to the inland territories of Poland and Lithuania which gravitated toward that sea: producers of raw materials themselves, these lands were criss-crossed with land routes over which Russian crude products were carried in transit. The growing populations and industry of Western Europe created a demand for imported grains, fiber crops and forest products. The turning point in the Baltic trade occurred in 1480. West European, and primarily Dutch, merchants broke the monopoly of the Hanseatic towns on the western Baltic coast and established direct trade dealings with the ports on the eastern coast of the Baltic, that is specifically with Gdańsk, Elbląg and Braniewo (Braunsberg), all ports that lay within Polish frontiers. The rising price of grain stimulated production. This was achieved primarily by expansion of large landed estates which were capable of raising surplus crops for export. As lords of large seignories East of the Łaba (Elbe) extended the area of land

* The mottoes placed at the head of each chapter are taken from the first large Polish biography of Copernicus, published by Szymon Starowolski in *Scriptorum Polonicorum Hecatontas*, Venice 1627.

they cultivated, they were able to affect an increase in the price for agricultural products at a rate incomparably faster than the craftsmen. Consequently, the 1520's witnessed a growing disparity in prices, especially as handicrafts organized along traditional lines could not keep pace with the trend toward industrialization. This called for investments which only the merchants with capital and large landowners were capable of making.

The far-flung territories of Poland and Lithuania were a slowly coalescing organism although they remained diversified in geography, nationality and religion. Next to the Poles, who were in the majority, the territories were inhabited by Lithuanians and Orthodox Ruthenians, by small groups of Armenians and Muslim Turks and by Jews and Karaims who professed Judaism. The Germans, largely assimilated by the Poles, formed the upper stratum of the large towns. Inhabitants of small towns were almost entirely Polish.

Distinct in nature within the borders of the Jagiellonian realm were the territories of Prussia that had been conquered by the Teutonic Order in the 13th century. Though partly colonized by Germany, they nevertheless contained a population that was predominantly Slav and Baltic Prussian. Geographically, Prussia occupied what were essentially two areas: the narrow and generally fertile strip of the Baltic littoral and the moraine hummocks of the lake region which still remain largely overgrown with forest. The wide valley of the Vistula (Wisła) and its delta cutting across the moraine ridges formed a distinct region. Because of its rich soil, the valley had been cultivated by man for many centuries, its area increased and protected by methods developed in Holland.

A dense network of strongholds covered the territory of Prussia in the 13th and the 14th century to offer protection against the threat represented by the oppressed native population. There was a total of 120 fortified Teutonic and episcopal castles with cities protected by these fortresses. In the period of insurrection the townsmen of large towns stormed and destroyed Teutonic castles while modernizing town fortifications.

In the 15th century a new conflict broke out between the settlers and the Teutonic Order. In 1454 the knights and townsmen of the territories of Teutonic Prussia rose, regardless of nationality, against the economic repressions and draconic rule imposed by the Teutonic Order. They appealed for help to the Kingdom of Poland with which the territories held by the religious order maintained close economic and cultural ties.

In effect the war, fought by mercenary troops and costing Prussian territories and Poland a great amount of money, exercised a significant influence upon world outlook. Traditional ecclesiastic institutions represented by the Teutonic Knights and their state – which lay within the domain of the Holy Roman Empire of the German People – were opposed by demands for the right of subjects to revolt against unjust authority, a concept developed by Polish theologians at the request of Prussian towns. The Kingdom of Poland, together with the territories of Prussia, surmounted the crisis evoked by the anathema deviously obtained from the corrupt and venal Curia Romana by the Order. The peace of 1466, concluded after a war that lasted thirteen years, sealed the doom of the Teutonic Order in Prussia with irrevocable finality. The Grand Master salvaged only a portion of his former domain and became a vassal of the Polish king.

The boundary line between the fief and the territories under the direct jurisdiction of the king was not drawn along a natural frontier. Consequently, the bishopric of Warmia constituted a virtually separate enclave connected with Royal Prussia only by a thin strip of coastline. The bishopric was almost entirely surrounded by territories that remained in the hands of the Teutonic Order.

The Thirteen Year War of 1454–1466, waged in a small area often with large military forces, brought economic ruin to the country. The population was reduced by more than half. It took over a hundred years to make up this loss. At the same time, however, the excellent economic opportunities, the unification of the territories, formerly under Teutonic rule, with the life-giving inland territories of a large state, the privileges obtained from the Polish king and the broad autonomy enjoyed resulted in quick reconstruction and reform of the structure. The father of Copernicus was swept from Cracow to Toruń on the wave of the 1458 migrations. Through marriage with Barbara Watzenrode, sister of canon Łukasz Watzenrode later bishop of Warmia, he entered the ranks of the Toruń patriciate.

Profits from Poland's transit trade to the Baltic were drawn by four large towns situated on the banks of the Vistula, the main artery by which Polish goods were transported. They were: Toruń, Chełmno and, at the mouth of the Vistula on the Baltic, Gdańsk and Elbląg. The patricians of these towns dominated the economy and the politics of the province, the most densely urbanized part of the monarchy. Due to the autonomy of the territories of Prussia, the economic benefits that accrued from the union with Poland, which provided the sole political defense against the decaying Holy Roman Empire and efforts of the Teutonic Knights to recover the lands they lost, and finally due to the attractions of Polish culture, the union of 1454, consolidated in common military effort, proved as natural as it was lasting.

With the economic stagnation of the territories of Eastern Germany and with the decline of the Hanseatic towns, elements of Netherlands culture appeared here together with the economic contacts. These were principally the *devotio moderna* and congregations of the lay and clerical Brothers of the Common Life who devoted themselves to a form of eduction of the young that became widespread among the wealthy townspeople. The ideals of respectability, simplicity and industry which they embraced complemented the instinct to count profits and to take risks, habits that were strongly entrenched in the merchant mind. The Brothers maintained ties with the world and devoted themselves to writing and teaching and also demonstrated an interest in Latin and classical literature. A school run by the Brothers is known to have existed in Chełmno at the time. By its outward reserve and seriousness of manner it represented a contrast with the exuberant activity of the Dominican and Franciscan Orders whose main interest lay in preaching. The art of the Netherlands found a ready market in Prussia and inspired local artists.

The diocese of Warmia, where with the exception of a few interruptions Copernicus lived for forty years, formed a separate entity within Royal Prussia. Its estates were divided between the episcopacy and the chapter. The one larger mercantile town was Braniewo with a few smaller towns economically subordinated to it. Hence Lidzbark with the bishop's residence, an impressive castle built in a square around an arcaded courtyard that may still be admired here, and Frombork, a port on the Baltic lagoon sprawling at the foot of a hill which supports a fortress-cathedral. The houses of canons (some of which have been preserved to our time) nestled outside the cathedral walls from which there spread a panoramic view extending beyond the bay rimmed with the dark of a forest and the narrow peninsula that separates the bay from the open sea. In the southern part of the episcopal principality, a full hundred kilometers from the sea, surrounded by lakes and forests and the gravel strewn moraines, the city of Olsztyn rises on a high cliff that overlooks the Łyna River. In the days of Copernicus the castle housing the administrators of the chapter lands dominated the city, today Olsztyn is a busy and fast growing capital of the province.

The agricultural region of Warmia skirting the sea was settled by Germans and Germanized Baltic Prussians while the southern, forested area (together with Olsztyn), was occupied by settlers who arrived from neighboring Poland. Polish settlers were brought in greatest numbers to take possession of villages abandoned after the destructive wars of 1454–1466 and 1520–1521. As administrator of the chapter lands, Copernicus was actively involved in bringing the Poles to Warmia. He visited the villages and kept a careful register of the peasants who settled here. Surrounded by Polish villages, the urban population of southern Warmia was swiftly Polonized. While the lands settled by Poles at an earlier time contained a fairly large number of wealthy and petty nobles and prosperous peasants, Warmia, being a poor and recolonized country, contained a negligable number of estates owned by the gentry and knights and a large group of less prosperous peasants.

2. The World of Copernicus

Left an orphan at a tender age, Nicolaus Copernicus spent his early childhood in Toruń, one of the principal towns of the Prussian territories. The town flourished and prospered after the Thirteen Year War. In the second half of the 15th century, however, Toruń allowed Gdańsk to take the lead.

Founded in 1231 by the Teutonic Knights on the banks of the Vistula in the Polish Province of Chełmno that directly bordered with Prussian territories, Toruń thrived principally on trade. It expanded in size when New Toruń, inhabited by tradesmen and artisans, was established nearby in 1280. In the 14th century the town waxed rich on commerce with Poland. The trade came abruptly to a halt in the early 15th century with the outbreak of war between the Teutonic Knights and Poland. The modern visitor may still be impressed by Toruń's splendid brick churches, the Old Town Hall, the town fortifications and the magnificent statuary and paintings.

The years of peace after 1466 mark a new phase in the development of Gothic art. The vertical accents of early Gothic architecture were now superseded by plain surfaces divided by horizontal lines. The parochial church of St. John the Baptist is perhaps the finest example of this style. The large edifice, expanded and enlarged in 1468–1472, rises above the city to this day. This is where we find the bronze baptismal font at which in the year 1474 a son of the Old Toruń merchant Copernicus was baptized and given the name of Nicolaus, the patron saint of merchants. Taking advantage of the support he received from his enterprising uncle, canon and from 1489 bishop of Warmia, the gifted boy was not enroled in the school ordinarily attended by merchants' sons. However, he received at his family home a working familiarity with arithmetic, the habit of using an astronomical calendar and a knowledge of money. The people of his time were excited by astronomical phenomena and internal urban affairs. Confined by restricted international horizons, they fully realized the value of thrift and combined an innate habit of taking risks with a respect for industry.

Destined for a church career and a higher education, Copernicus was quite probably sent to Chełmno where the *studium particulare* was established on a higher academic standard than the parochial school of Toruń.

Chełmno was located a day's journey down the Vistula from Toruń. Although it was one of the major Prussian towns, it sank into economic stagnation after the war. However, the old school tradition (steps had been taken in the 14th century to set up a Prussian university here) still remained alive. The Brothers of Common Life offered the students other stimuli and moral concepts than the idea of getting rich at all cost which was the moving spirit of Copernicus' native town.

In the autumn of 1491, Copernicus then eighteen years of age, entered the University of Cracow instead of one of the Baltic universities at Rostock or Greifswald. His decision was motivated by family tradition and the reputation the Polish academy enjoyed. It was considered one of the finest universities of Central Europe at the time.

Cracow was not only the capital of a large power but also a commercial metropolis, a market town of an agricultural region and administrative center of a mining district. Its population, larger than that of any town in Prussian territories, was multinational in character. Next to the German patricians, quickly assimilated by the predominantly Polish population, came the Italians who also made their home here. In some ways the magnates dominated the townsmen.

The political horizons were circumscribed by the territories over which the block of four countries ruled by the Jagiellonians, extended. The boundaries of the block, composed of Poland, Lithuania, Bohemia and Hungary, extended in the south to the Adriatic and the Black Sea. The Baltic issue receded into the background as the Turkish danger grew ominous. With the loss of the Black Sea ports to the Turks (Kaffa-Feodosiya in the Crimea in 1475 and Chilia and Belgorod Dnestrovski (Alba Iulia) in 1484), the countries began to mount the disastrous retaliatory expedition of 1497 to the mouth of the Danube.

Cracow with its white stone structures in no way resembled the red brick towns of the Prussian territories. Gothic was entering upon a new phase of development when Copernicus arrived there. The new style incorporated window tracery, flamboyant fan vaults that came to central Europe by way of Saxony, Silesia and Upper Hungary (Slovakia).

The new form of architecture was preceded by great works of painting and altar sculptures. The young man from Toruń and Chełmno was awed by the magnificence of group compositions of the late Gothic and by the remarkable miniatures he saw. (Toruń was not to received the great 'Descent from the Cross' until 1495).

Church interiors were furnished with pictures, tapestries, sculptures, decorative gold leaf ornamentation designed generally with figural though also with floral and geometric patterns. As historical sense in the modern meaning of the term was alien to the genius of that age, biblical scenes were placed in contemporary settings and ancient figures mingled with living persons and saints popular at the time. The trend fostered by the new intellectual attitude that took hold of the country, that is by interest in the discovery of the world and man, provides a fairly good picture of 15th century man, of his appearance, the manner in which he dressed and the kinds of work he performed as seen in the various social, anthropological and emotional aspects. The limited range of symbolic colors and the hieratic figures were now more diversified, the shades of light and the very shape of life were depicted with greater veracity. All these developments served to stimulate public imagination.

While custom and technical difficulties weighed heavily on the monumental, poster-like altar paintings, there were fewer obstacles to innovation in the elite art of the miniature which flourished in Cracow in the 16th century. Many fine illuminated volumes of liturgy and private prayer books were brought out at the time. Private and individual experiences of men of breeding and taste began to play a larger role in the illuminations as did the surrounding secular world, with its fascinating sweep of landscapes and touching detail. The most remarkable, and this a secular, work produced at the time was the *Codex picturatus*, commissioned by the town clerk of records Baltazar Behem. The book is a collection of the statutes of the Cracow guilds illustrated with 28 miniatures. They form a series of pictorial sonnets depicting artisans at their various occupations. Workshop interiors, landscapes, symbolizing and rhapsodizing snowy winter and hot summer, form a unique encyclopedia of urban life inscribed in luminous gold and subtle shadings of color, ranging from black and white to lemon yellow and blue. We see the same range of color in the pictures painted by Vermeer 150 years later.

Painting and sculpture, with the glints of gold, the glowing purity of color, the drama of narrative and the sacred nature of the subjects presented, had a powerful and lasting effect upon human emotion and gave impetus to intellectual development.

The finest expression of the art of Cracow were the large painted and carved triptychs and polyptychs enclosed within the sculptured frames of Gothic architecture. The triptych, with its side panels thrown open during the service, was intended to dazzle the people with its opulence. Since the triptych was subordinated to the liturgical rites performed in front of the altar, consequently, the style tended to be conservative and hieratic in form. The backgrounds for the figures and scenes from the bible and from the lives of saints were invariably in gold. The less luxurious exteriors of the side panels (visible when the triptych was closed) were executed by young and very often highly inventive artists. The scenes were free of the lavish use of gold and represented the ordinary world. The first large Cracow triptych, found in the Dominican Church (1477), marks the beginning of realism and the wider use of a rich profusion of flora and fauna in the landscapes that were clearly reminiscent in hue of the countryside of Cracow. Events that took place in distant lands and in remote ages were symbolically depicted as occuring now and here in the country lanes and in the presence of neighbors, shepherds and their herds, artisans and merchants. Birds sing in the trees, a mole can be glimpsed through the flowers and grass as it burrows in the fresh plowed earth and an angel, resembling a large insect, shoots across the star-studded firmament like a falling star.

The greatest masterpiece of the European art produced in that century, one whose completion was witnessed by Copernicus, was the polyptych that took the German sculptor Wit Stosz (Veit Stoss) a full twelve years to carve, gild and paint. More gold was used in the adornment of the altar, founded with contributions made by the townsmen of Cracow – most of whom were Poles, than in any other altar of that day in Europe. But it also must have provided the strongest emotional experience. The center panel represents the 'Last Sleep of the Virgin Mary'. As the Apostles cluster around her in great agitation, her body sinks gently to the ground. The agitation is expressed by concern and indignation painted on their faces, by the wringing of hands, the gnarled legs and the heavy gilt robes billowing in the wind. Each disciple reacts differently to the death of the Virgin (or is it a reaction to the death of an age and order in which the artist lived?). The scenes on the side panel done in polychrome low relief relate not only the story of the life of the Virgin Mary and the Passion of Her Son – a tale dear to the heart of the people – but also recount the epic, or rather make an encyclopedic enumeration of people and objects, of the sins and virtues of the dying middle ages.

The triumph Stoss celebrated with the altar he made for the townsmen of Cracow opened to him the portals of the royal cathedral at Wawel. Stoss produced another masterpiece here. This time it was a mausoleum for Casimir Jagiellończyk who died in 1492. The movingly realistic death mask, the king's figure in red veined marble twisting in the agony of death lies on a tomb engraved with figures of traditional mourners which contrast with the sumptuous canopy of intersecting late Gothic arches that represent the firmament.

In Cracow the great sculptor spent the happiest and most productive years of his life (1477–1496). From his tragic Nuremberg period comes the bronze plaque of Philip Kallimachos (born Filipo Buonoccorsi), Italian historian, humanist and political advisor to the Jagiellonians, owner of a townhouse that stood next to the home of the Copernicus family in the Old Town Market Square of Toruń. The bronze plaque was cast in Nuremberg for Poland by Peter Vischer.

When Stoss departed from Cracow, he left behind him pupils and imitators who were active in a large area that stretched from Slovakia as far as Warmia. The many remarkable works they produced gave a true picture of Poland and Poles of the early 16th century. Fine paintings continued to be produced in Cracow. Notable among these was Marcin Czarny's 'Last Sleep of the Virgin Mary' depicted on one of the panels of the Bodzentyn triptych of 1508 which also contains a full length portrait of its founder, the Cracow bishop, Jan Konarski.

The democratic, black and white art of printing developed with the support of the University. The faculty of liberal arts at the Cracow University where Copernicus spent four years as a student had a fine tradition which dated back to the first half of the 15th century. Students, these chiefly from Central Europe, continued to flock to this still very vigorous academic community. Lively contacts were maintained with Italian humanism through a course in the masterpieces of literature. We might mention the fact that in 1492 ten lectures were given on the subject of classical literature, namely the works of Cicero, Ovid and Virgil (his *Bucolics*, the *Aeneid* and the *Georgics*). Jean Buridan's interpretation of Aristotelian physics was expounded in the natural sciences, hence principally the theory of momentum which represented an approach to the motion of bodies that laid the foundations for the development of modern physics and astronomy. Buridan also formulated the proposition regarding direct observations of natural phenomena. Mathematics and astronomy attained a high level of development at the university of Cracow. It seems quite likely that the nephew of the bishop and politician Watzenrode did have some connection with the extramural 'second circle' of intellectual culture where itinerant humanists occasionally mixed with bishops, politicians and diplomats.

Although Copernicus did not obtain a degree at Cracow, there is no doubt that he did acquire an education there. His knowledge of mathematics and first observations made clear to him the contradictions inherent in the theory of the geocentric system of the universe. He left for Italy in 1496 bearing with him this rankling problem and a thorough grounding for further studies.

He arrived after the 1495–1496 revolution in military technology. This was a time when in the service of their countries scholars and artists were feverishly bent on discovering new technological solutions which would counteract the destructive force of artillery and the strong defenses of fortifications. They looked for new uses to which the mobility of artillery could be put in battles waged in the field. Spanish and French commanders developed new infantry tactics that were later put to the test in the battlefields of Italy. We do not know what impression the ancient monuments and the discovery of perspective represented in masterpieces painted by contemporary artists that he saw in Bologna, Ferrara, Padua and the Rome of Alexander VI made on the student from the kingdom of Poland who, outside the regular course of study of medicine and law (the official purpose of his journey to Italy), devoted himself assiduously to the computation and the ever more scrupulous checking of data gathered from the observation of 'the heavenly spheres'.

Having obtained a doctorate in canon law, Copernicus hastily returned to Warmia where he was to remain to the end of his days. Until his uncle's death in 1512, he lived at the castle of Lidzbark and accompanied the bishop to Cracow on two occasions and most likely once to Poznań. In Cracow he must have surely observed the changes in the architectural panorama, the new Gothic fan vaults of the Collegium Maius completed in 1497 and the first works of the king's Italian architects representing the Tuscan renaissance style in its purest form. He must have also seen the latest works of Veit Stoss and the sculptures produced in abundance by his pupils as well as the first renaissance works of Florentine artists. Moreover, he could observe the technical and economic progress of the country and of its capital and renew old friendships, principally perhaps with Bernard Wapowski, a remarkable cartographer and historian; Copernicus corresponded with him in later years. Moreover, nowhere but in Cracow could Copernicus publish his first book, the humanistic *Letters* of Theophilactos Simocatta he translated into Latin from the Greek.

He was in Cracow too early to meet the illustrious humanists born about the year 1500 (except perhaps for Stanislaus Hosius, canon of Warmia from 1538 and later cardinal). However, he must have heard of issues relating to the internal policies of the Jagiellonian states, principally of the movement of the wealthier gentry who as they grew rich threw off the supremacy of the magnates and sought economic privileges. However, Copernicus remained most vitally interested in Prussian affairs.

He accompanied his uncle frequently to the provincial diets of Prussia and continued to attend them even after 1512 (the year of his uncle's death) as delegate of the Warmia chapter. Hence, he traveled occasionally to Gdańsk, Toruń, Grudziądz, Elbląg. Most often, however, he journeyed to the former Teutonic capital of Malbork. Its castle, at one time the residence of the Grand Master, was one of the finest buildings in Europe. Its massive bulk still looms over the wide currents of the Nogat River.

Oriented in the arcana of politics and economy, Copernicus commanded due respect of the magnates and members of the gentry and clergy he encountered at the provincial diets and other meetings as well as of the representatives of large and small towns. For this was an age when even the ideology of knighthood proclaimed that the two pillars without which the world would be reduced to chaos are 'chevalerie et science, qui moult bien conviennent ensemble'. The title of doctor was accorded the same privileges as the knights enjoyed.

His immediate homeland, Royal Prussia, changed its aspect in its own peculiar fashion, although the light and shadow effects of the fan vaults and the softer contours of the renaissance also made their appearance in Warmia. The pointed arches of red brick became rounded under the influence of Netherlands architects who arrived by way of the Baltic and of the Saxon artists who came by overland routes that led from the Central European commercial metropolis of Leipzig. The cares and concerns of the Prussian gentry, whose spokesman was the Italian educated Fabian of Losainen (1512-1523), successor of Bishop Łukasz Watzenrode, impressed the minds of the intellectuals congregated in the Frombork chapter. The interests of the powerful Gdańsk patricians were represented by the next bishop, the authoritarian and ruthless Maurycy Ferber (1523–1532), while the grand design of the royal policies of Cracow were personified by Jan Dantyszek, poet and courtier, for many years envoy to the court of Charles V at Valladolid, later correspondent and patron of Erasmus of Rotterdam. Jan Dantyszek was bishop of Warmia in 1537–1548.

The chapter, to which Copernicus was appointed in 1495, differed from other corporations of this type by the fact that it was composed almost exclusively of townsmen. The remarkable majority of the canons had a higher education, although only few of them were ordained to the higher priesthood. Among them were such leading humanists as the gifted poet and diplomat Dantyszek and Tiedemann Giese, correspondent of Erasmus of Rotterdam and Philip Melanchthon, close friend of Copernicus, from 1538 bishop of Chełmno and from 1549 bishop of Warmia. Usually only half of the canons remained at the chapter for others were busy collecting revenues from benefices scattered all over the country, performing duties in the service of the king and studying in foreign lands. Consequently, although the tradition of common life was still practiced by the canons of Frombork, they were nevertheless 'landlords of Frombork' rather than clerics; they were principally concerned with the economic needs of their seigniories and the administration of the property held in common by the chapter. Copernicus had friends (although there were also some intrigants) but he was in the main a recluse as his chief passion, an interest he pursued with great ardor in the time he had left from his economic and political duties, was computation and observation. He carried on this work without interruption throughout the war. The revolutions of celestial spheres continued in their course without regard to the fortune of war on earth. In the light of these facts, his friendship with Tiedemann Giese was unique. Copernicus urged Giese, for many years the only priest in the chapter, to take issue with Lutheranism, in the Erasmian spirit quite likely, Giese became interested in astronomy and as bishop of Chełmno prevailed upon Copernicus to write and publish a complete study and not only the tables as the astronomer was inclined to do.

The life of these men centered around a unique cathedral fortress whose walls, constructed in an irregular pentagon and reinforced with towers, enclosed a levelled area on a steep hill. The northeastern part of the area is occupied by what is perhaps the grandest Gothic church east of the Vistula. The long body with towers topped by pinnacles, built in 1329–1388, follows the ground plans common in Cistercian church architecture. The unadorned side walls pierced with windows that offer the only relief to their plainness contrast with the ornate facade, its gables divided by elongated false windows and an arcade. Elements of Götland sandstone are covered with intricate carvings. The opulent interior, its length together with the choir closed in a straight line (86 meters), is enhanced by the flamboyance of the rib vaults that surmount the entire building. The main accent was provided by the large folding altar (now in the north aisle). Founded by bishop Watzenrode in 1504, its erection was witnessed by Copernicus. The sensuous figure of Madonna and Child enclosed in the gold interior is the work of a Toruń sculptor.

Of the most immediate concern to Warmia and Prussia, hence also to Poland, was the proximity of the Teutonic Knights. In the early 15th century the Teutonic Order of the Virgin Mary could still count on hundreds of brethren from religious orders and on thousands of crusaders from nearly all the European countries. To an appeal to defend the Christian commonwealth they responded by arming at great expense and arriving in force. The Teunotic Order waged the Thirteen Year War of 1454–1466 with mercenaries alone, these mostly foot soldiers. The contrast between ideology and practice became most conspicuous in the early 16th century when the objective was no more than the preservation in an alien land 'a hospital of the Germanic knighthood', which having lost basis for existence in its native land now desired to hold on to the revenues (facilities) in the colonies. In complete disregard for the conditions laid down by the peace of 1466, the Order did not admit Polish knights into its ranks. It sought not only to dissolve the bonds of vassalage but also to recover the whole of Prussian territories.

By appointing the most powerful princes of the Empire as Grand Masters, hence Frederick of Saxony in 1510–1517 and then in 1517–1525 Albrecht Hohenzollern, a nephew of the elector of Brandenburg, the Teutonic Knights counted on German help and identified their cause with the aspirations of German princes which were motivated by a desire for the disintegration of the empire, particularism and absolutism. These ambitions represented a threat to the interest of the Prussian population and to their liberties and prosperity which were conditional upon the adherence of Prussian territories to the Polish economy. They represented a danger to the interests of Poland, to the country's access to the sea and its ties with the western markets and stood in direct opposition to the political ideology of democratization, the broadening of popular participation in the government of the state, of a state that extended over a vast area where a diversity of economic factors interacted.

An urgent question, one of particular concern to Prussia, was that of money – the bad money of the Order which drove out good Polish money – and that of the immediate safety of Prussia and particularly of Warmia which found itself surrounded by lands held by the Teutonic Order.

Even after the death of his statesman uncle in 1512, when Copernicus finally established himself at the cononry of Frombork, he had to interest himself in the broad political forces at play in which the conflict with the Knights was an important factor. He also had to devote himself in particular to the security of Warmia and to the region's predominantly agricultural economy when the chapter appointed him administrator of its property. Thus, as a physician he prescribed cures for his colleagues and as a jurist and canon he sought remedies for curing the ailing monetary system. As administrator of the chapter's estates, he brought settlers to abandoned peasant fields and as mathematician and engineer he oversaw and organized the construction of modern artillery towers at the castles of Lidzbark, Reszel, Olsztyn and perhaps even of Frombork.

With the approach of the next conflict (the 1519–1521 war with the Teutonic Order) and with the flight of the canons from Frombork when that town was seized by the Grand Master, Olsztyn, administered at the time by Copernicus, became – next to episcopal Lidzbark – the main point of resistance. The burden of the diplomatic mission to the Grand Master, concern for the chapter's treasury and for the fortifications of the castle and town devolved upon Copernicus. It was also Copernicus who took charge of the garrison of Polish soldiers and it was Copernicus who bore the responsibility for the safety of Olsztyn when the Grand Master led an audacious raid against the town.

The years of war followed by years of truce finally ended with the 1525 treaty of Cracow and homage paid (to the Polish king) by Albrecht Hohenzollern who cast off his clerical robes and became a lay prince in Prussia. These years were also marked by upheavals of a different nature: the Reformation and a wave of bloody social movements against the magnates of the world – both lay and spiritual. Social tensions, greatest between the commonalty and the patriciate of Prussian towns, prepared the ground for the acceptance of the Reformation which spread quickly by means of print and sermons. The Reformation provided an ideological framework for the grievances put forth by the townspeople of Gdańsk, Elbląg and Toruń who had ardently espoused the principles of a plain church. These demands ended with checks to royal power and to the abuses of the partriciate by appointing a representative body of commoners to control the town economy. The social discontent of the town commonalty and poor manifested itself in riots and even in bloody insurrections which reached a culmination in 1525 when revolts broke out in Elbląg, Gdańsk and Toruń as well as in Warmian Braniewo, just ten kilometers from Frombork.

More sinister, however, seemed the peasant revolt that broke out in 1525 in Sambia, a Teutonic state just across the border from Warmia. The revolt was directed against growing oppression which, oddly enough, coincided with Albrecht Hohenzollern's acceptance of the Reformation against the wishes of many of the brethren of the Teutonic Order. In crushing the peasant rebellion, the protestant prince secured the support of Polish nobles and of the Roman Catholic bishop of Warmia, Ferber.

The spread of the new religion was enjoined by the king of Poland, Sigismund the Old, at the penalty of death, although the monarch recognized the break between the papacy and Albrecht Hohenzollern and his petty state.

Of Copernicus we know only that he remained loyal to the Church to the end of his days in the same way as the great humanist and critic of abuses within the Church, Erasmus of Rotterdam, who was five years his senior. One may wonder why. Perhaps the reason was that he wished to keep the canonry which provided security and a place for his scientific work. Besides even the excommunication of Luther in 1520 did not necessarily signify the collapse of Christianity to people accustomed to the excommunications of the late middle ages. Then there was also the irenic hope for a reconciliation. As the majority of humanists, so Copernicus was doubtlessly a loyal Christian. He must have looked upon the Church, cleansed of its errors, as upon a bulwark of social order, which was so dear to the partricians from whom he sprang, and as a guarantee of the unity of the intellectual world to which he belonged. As Erasmus, so Copernicus did not find *His Road to Damascus*. Moreover, as he grew older, he began to concentrate more and more on the completion of his life's work, and his energies were beginning to fail him.

NICOLAUS COPERNICUS – HIS LIFE AND WORK

JERZY DOBRZYCKI

Faculty of the History of Science and Technology of the Polish Academy of Sciences

> Ut Hercules Atlante laborante caelum sustinuit, ita ille... motuum caeli doctrinam discipulis suis restauravit

1

In the 15th century mathematical astronomy attained a point, at least in the works of its leading exponents, where it was no longer possible to avoid making an effort to provide an answer to two basic dilemmas. The first of these was the question of the feasibility of astronomical theories, a problem inherited from ancient natural philosophy. The controversy on the conventionality of scientific theory was to continue into modern times although the substance of that controversy was to undergo a certain change. The second problem pertained to the credibility of established authorities and was rooted in the contradiction that lay between old theories and modern data based on observation.

The Aristotelian astronomical system was an inseparable component of his pictures of the world. According to Aristotle the world was built in concentric crystal spheres that carried the Sun and planets in one eternal and perfect circular motion around the Earth which rested immovably in the center of the universe. This system, expressed by mathematical formulas as early as the 4th century B.C., became, in the course of the centuries in which the Aristotelian doctrine remained the dominant one, strongly entrenched in the consciousness of men of all levels of education as a system which accorded with natural philosophy as a whole and with elementary observations that lent support to the concept of the spherical structure of the firmament.

However, the system of concentric spheres failed to provide a satisfactory explanation to the irregularities noted in the motion of the planets, due to the fact that their orbits are not circular but elliptical; it also failed to explain the changing distance of planets from the Earth – manifest in the shifting intensity of their light, which fact is not reconcilable with the concept of the constant distance of concentric planetary spheres. Astronomy made an effort to establish harmony between the geometric model and observed facts by imaginative constructions. In the 2nd century B.C., Hipparchus' acceptance that the center of the Sun's orbit lay beyond the Earth stood in satisfactorily for Kepler's elliptical motion. From the geometric works of Apollonius of Perga (3rd century B.C.), Ptolemy elaborated the idea of the epicycle, that is a circle in which a planet moves at a uniform velocity; it is the center of this which is carried, also at a uniform velocity, around the circumference of the main circle (deferent) of the orbit. The motion of a planet along an epicycle represented in geometric terms the actual motion of the Earth. We discover in Ptolemy's *Almagest*, which contains the sum total of what the ancients accomplished in the area of geometric descriptions of the universe, yet another deviation from the pure Aristotelian doctrine. This took the form of the construction of what is known as the equant. According to this construction the angular velocity of the center of the epicycle around the deferent remained uniform not in relation to the center of the deferent but in relation to the 'level point' or 'point of equality', that is to the equant, which occupied a different position than did the center of the deferent and the Earth.

The discrepancy between astronomical practice, which made use of the system of circular motions, and natural philosophy, where the hierarchy of concentric planetary spheres was in force, remained over many long centuries the central unresolved problem in astronomy. Although efforts had been made since the days of Ptolemy to reach a compromise by combining the two systems, they did not produce satisfactory results; solution was sought by accepting the theorems of mathematical astronomy as orthodoxical in character. The Platonic postulate explaining phenomena (by reduction to elementary circular motion) became the formula for 'saving appearances', a term applied to the arbitrary character of mathematical models reduced to the role of arithmetical formulas. That point of view was to exercise an important influence on the reaction to the heliocentric theory.

The second problem rose into sharper focus together with the growing interest in the original classical texts. And so, the spherical system of reference, with the positions of celestial bodies designated on it, was set forth in the *Almagest* by numerical data which were sufficiently inexact to make it impossible to reconcile them with observations conducted over the subsequent centuries. In the end, medieval astronomy had to stop the practice of applying Ptolemy's (numerically) erroneous theory 'of the motion of the eighth sphere' and to replace or supplement it with more complex schemes. Compendiums of astronomy (the *Theoricae*) of the 13th to 15th century, left this question unresolved as well, taking note only of the opposing views held on the subject by Ptolemy and by later authors. Yet the theory of the 'motion of the eighth sphere', which described the phenomena of precession, was important to basic astronomical measurements. Furthermore, in view of the advances made in the art of observation and in view of the need for a reformation of the calendar in the 15th and 16th century, this theory became the major problem facing mathematical astronomy. Premises for taking up these questions, that re-emerged in the 15th century with the growing professional competence of astronomers, were reflected most sharply in the first half of the century by the activity of the 'Vienese school of astronomy' which produced authors of the most important works of that period, namely Georg Peurbach and his pupil and associate Johann Müller of Koenigsberg in Franconia, known by his Latin name Regiomontanus. Both scienists died before they were able to complete their plans to purge the text of the *Almagest* of the corruptions that successive translations from the Arabic carried out over the centuries had introduced into the work. They did, however, succeed to enclose the sum of astronomical knowledge of their age in such works as Peurbach's *Theories of Planets* and the *Epitome in Almagestum* (a summary of the *Almagest*) by Peurbach and Regiomontanus. Mathematics reached new heights of competence in the mathematical writings of Regiomontanus.

Toward the end of the century, the Academy of Cracow became the most active center of astronomical science. The Cracow school of astronomy developed around the two permanent faculties of astrology at the academy congregating a whole galaxy of scientists. Notable among these was Marcin Bylica, who when he travelled abroad had worked with Regiomontanus, and Wojciech of Brudzewo, author of the astronomical tables and commentaries to Peurbach's work.

The growing importance of the Cracow astronomical school in the second half of the 15th century constitutes a significant element in the history of European science. Most important, however, is the fact that it provided the foundations from which Nicolaus Copernicus developed his creative interest and evolved his research goals.

The founder of modern astronomy, Copernicus, was born in Toruń on February 19, 1473 to a family of burghers, a social formation that played a major role in the political and economic life of the age. The Copernicus family came from Lower Silesia (the village of Kopernik, now Koperniki near Nysa). The astronomer's father, the merchant Nicolaus Copernicus, was descended from a branch of the family that lived in Cracow. Trading in copper, he was in close touch with the Prussian towns and played a part in activities that led to the founding of the Prussian Union of 1454. A few years later, he settled in Toruń and married the daughter of the merchant Watzenrode, Barbara, the astronomer's mother. The Watzenrode family, also originally from Silesia, settled in Toruń a couple of generations earlier and by now belonged to the town's patriciate. Barbara's father, hence the grandfather of Nicolaus Copernicus, held the office of the elder of the (Old Town) court magistrate in Toruń. He was also active in the Prussian Union and took part in the fighting against the Teutonic Order. The brother of the astronomer's mother, Łukasz Watzenrode (the younger, 1447–1512), studied in Cracow, Cologne and Bologna and filled a number of church posts. In 1489 he was appointed bishop of Warmia. Łukasz Watzenrode was one of the most able politicians of his age and, having won the confidence of the Jagiellonian rulers, took part in shaping Poland's policies toward the Teutonic state. After the death of the father of Copernicus (about 1483), Łukasz, as guardian of his sister and her family, was to play a decisive role in shaping the future of his nephew Nicolaus.

Copernicus spent his pre-school years at home, at first in the house were he was born at St. Anna Street (now Kopernika Street) and from 1480 at a house that stood in the Old Town Market Square. It is still a matter of debate what school he attended. It seems quite probable that he went to the municipal school of St. John's Church. The school had a certain reputation in the instruction of astronomy and it was run for a few years by the future bishop of Warmia, Łukasz Watzenrode. Copernicus might have continued his education at the Brothers of Common Life school in Chełmno.

Following his uncle's advice and example Copernicus enrolled at the University of Cracow in 1491. For the four years he remained at the University he devoted himself to the study of the liberal arts without, however, winning an academic degree. The liberal arts course formed the basis of university education at the time. The curriculum included a broad range of instruction in the mathematical sciences. As regards astronomy, Copernicus was able to attend lectures on such objects as the Sphere *(Sphaera)* by John Holywood (elements of cosmography), the *New Theories of the Planets* (*Theoricae novae planetarum*) by Peurbach with commentaries by Wojciech of Brudzewo (descriptive exposition of theoretical astronomy) as well as introductory courses in the application of astronomical tables: the Cracow version of *Tabulae resolutae*, Peurbach's eclipse tables and the Regiomontanus ephemerides. The crowning point of lectures in astronomy was a course in astrology based on the *Tetrabiblos* of Ptomely and on the astrological treatise of Haly Abenragel. The names of instructors whose lectures Copernicus attended have been established with a greater or lesser degree of accuracy thanks largely to the researches of L. A. Birkenmajer. Hence, it is known that Bernard of Biskupie and Wojciech Krypa of Szamotuły lectured on *Tabulae eclipsium* and on the *Tetrabiblos* of Ptomely at the time Copernicus was a student at the University. Copernicus may have also attended lectures conducted by Wojciech of Pniewy, Szymon of Sierpc, Michał of Wrocław (astronomy) and Jan Głogowczyk (geography). In 1493 Wojciech of Brudzewo commented on *De caelo*, an important philosophical and natural science treatise of Aristotle.

Clearly the quality of knowledge contained in the series of courses on astronomy offered at the university should not be overestimated. *Theoricae novae* gave a simplified description of orbital mechanisms without, however, employing the highly elaborated mathematical apparatus found in Ptolemy or methods that led from observation to geometric generalization. It was possible to learn from the information provided by the astronomical tables the rules of arithmetic applicable ad hoc, particularly in astrology. No attempt was made, however, to relate these tables to the theories they represented. But mastery of this mathematical material and the discovery of contradictions within the prevailing astronomical doctrines were both essential steps in the views of Copernicus. Equally important was the atmosphere of the Cracow center of scholarship, renowned in Europe for its teachings in mathematics, for it provided the stimulus to his later activity. Through his direct contacts with the Cracow astronomers Copernicus was able to gain insights into more advanced problems in geocentric astronomy. According to established tradition Copernicus is said to have also attended extramural lectures given by Wojciech of Brudzewo. Quite likely, it was through private instruction obtained in a similar manner that Copernicus acquired the skill of conducting astronomical observations.

Thus, it might be said that the entire Copernican opus grew organically out of the tradition of the 'Cracow school of astronomy', a fact of which the astronomer was fully aware. An account of 1542 makes reference to Copernicus in this manner:

> whatever he has already written fully admirably in mathematical subjects and whatever he intends to publish, he has taken from this our university as from a wellspring; which fact he not only does not deny... but admits himself that he owes everything to our Academy.

Copernicus left Cracow in 1495 departing for Frombork to take up his duties as canon of the Warmia chapter which had its residence beside the Frombork Cathedral. Bishop Łukasz Watzenrode, who was no doubt instrumental in securing the nomination for his nephew, had thus gained a trusted ally in the chapter. On the other hand,

Copernicus with the guarantee of a prebend attached to the office of canon was now secure from financial worry. It was not necessary to be an ordained priest in order to qualify for the office of canon and there is no evidence that Copernicus had ever taken the holy orders. The astronomer remained in Warmia but a short time, for the very next year in 1496 he set out, again following the example of his uncle Łukasz Watzenrode and with his support, to study law at the University of Bologna, a famous school of jurisprudence. The law studies, which he was to continue until 1500, did not absorb his attention wholly and Copernicus was able to pursue his interests in the humanities and in astronomy, his life's passion. Already in the early part of his Bologna sojourn Copernicus began to collaborate with Professor of Astronomy Dominico Maria Novara. Significant is the substance of the observation made by Copernicus on March 9, 1497, quite likely together with Novara. He was 'not so much (his) pupil as (his) assistant and witness of observation'*. The event was the occultation of Aldebaran, a star of the first magnitude in Taurus, by the Moon. It offered the opportunity to check the Ptolemaic theory of the motion of the Moon. The theory was encumbered with a fundamental error regarding the changes in the distance of the satellite from the Earth. It was disclosed through the observation made in Bologna, Copernicus was to draw upon it later in his major work, that the parallax of the Moon (hence also its distance) in its quadratures does not differ from the parallax of the full Moon. This discovery contradicts the model given by Ptolemy. Yet disclosure of an error in an old theory of the Moon did not by itself overturn the geocentric system of the world. It did, however, represent an important step in the creative development of Copernicus and constituted proof that he had entered upon a new phase in this development. The new phase was marked by a conscious questioning of recognized authorities on the basis of observed evidence that had been rationally and relevantly selected. We read further in Rheticus' account that Copernicus also began to observe the stars at this time. These observations were necessary to the formulation of the theory explaining the still unresolved enigma of the the 'motion of the eighth sphere'.

After a brief stay in Rome (1500), which might have been connected with legal and office practice in the Curia, Copernicus returned to Poland the following year to once again obtain permission from the chapter to continue his studies. This time it was to be medicine at Padua, famous for the high standard of instruction in that subject. Two years at the University of Padua prepared Copernicus for the medical profession (he was to practice medicine up to the final years of his life). Furthermore, at Padua he was introduced to a circle actively devoted to humanistic philosophy and to the study of ancient culture. It was here that Copernicus learned Greek and read extensively in ancient literatures. By learning Greek he not only attained the ideal of a humanistic education but fulfilled the indispensable condition by reading the works of ancient writers in the original, unmarred by translation. His concern for restoration of the correct meaning of the texts, particularly important in matters of chronology, is

* According to the account given by Rheticus in *Narratio prima*, Gdańsk 1540.

documented by the numerous corrections made in his hand in the books he owned.

The constructive phase of 'the Copernican revolution' took place at Padua at the latest. It included: search for geometrical solutions which when applied to astronomy would fulfil the postulate of the homogeneous and harmonious nature of the universe and which would liberate it from inconsistencies such as the equant*. By the time Copernicus completed his Italian education, crowned in 1503 with a doctorate in canon law received at Ferrara, the new picture of the world, which accorded with the above postulates and which was supported by arguments based on its ancient fore-runners of which he learned through his study of ancient literature, was already firmly established in his consciousness. The detailed description of the new theory must have been developed at least in part by then for a few years later Copernicus was able to show to a small group of trusted friends a draft of the heliocentric theory set down in writing.

Upon his return from Italy Nicolaus Copernicus made his home in the Lidzbark residence of the bishop of Warmia. The chapter records indicate that because of his skill in medicine he was to perform the role of the bishop's private physician. In fact, however, he was among Łukasz Watzenrode's most intimate collaborators in the administration of the diocese and in the political activity into which Bishop Watzenrode was drawn by the constant conflict with the Teutonic Order. Copernicus played a part in this activity as the bishop's trusted collaborator. He accompanied the bishop to a number of political congresses and sessions of the Crown and to the Prussian Sejm (diet). In connection with the border dispute with the Teutonic Knights, Copernicus also drew a map of Royal Prussia (circa 1510)**. He was to continue Bishop Watzenrode's political line later when he settled permanently in Frombork shortly, it is assumed, before the death of Łukasz Watzenrode in 1512. From 1510 Copernicus performed an administrative function in the chapter (as chancellor and estate inspector).

While still residing at the side of Bishop Watzenrode, Copernicus made his debut as an author revealing his interests in philology. His Latin translation of letters of morals (*Epistolae morales rurales et amatoriae*) of the Byzantine writer Theophilactos of Simocatta was printed by Jan Haller of Cracow in 1509. The letters were dedicated to Bishop Watzenrode and the text was preceded by a verse written by Wawrzyniec Korwin, a friend of Copernicus and his teacher when he was a student in Cracow.

With his Lidzbark period we primarily relate an event which was to have a momentous influence on the history of science. This was the writing of an outline of the principles of astronomy. The short treatise was the first exposition of heliocentric astronomy.

We do not know for certain the exact circumstances or the date of the writing of the

* Describing in the introduction to *De revolutionibus* the reasons why he undertook this work, Copernicus critizes his predecessors who 'failed to discover or to educe from them what was most important, that is a system of the universe and the established order of its parts'.

** The map has been lost. The interest of Copernicus in cartography is supported by information from later years of his life.

treatise nor do we know its original title. The treatise preserved in copies made at a later date bears a Latin title of doubtful authenticity. It reads: *Nicolai Copernici de hypothesibus motuum coelestium a se constitutis commentariolus.*

There is no date affixed to the treatise, It was possible to establish the approximate dates within which it could have been written thanks to the studies conducted by L. A. Birkenmajer. The terminus ad quem is determined by the fact that the 1514 inventory of the library of Maciej of Miechów mentions the short treatise 'maintaining that the Earth moves and the Sun remains in a state of rest'. The affiliations between the text of *Commentariolus* and *Epitome in Almagestum Ptolemaei* by Regiomontanus, published in 1496, and reference in the manuscript to the astronomical tables of 1502 would enable us to establish the terminus post quem. Finally, we find allusions to *Commentariolus* in the verse by Wawrzyniec Korwin mentioned above. Referring to Copernicus, the author writes,

he "knew how to explain the hidden causes of phenomena on principles worthy of admiration."

Korwin, a resident of Wrocław, stayed in Toruń in 1506–1508. This fact allows us to assume that *Commentariolus* was written about 1507. The work was not meant for publication. Only a small number of copies was made and these were distributed above all (if not exclusively) among members of the Cracow circle.

Commentariolus begins with a reminder of the 'primary principles of absolute (i.e. uniform motion) and with a criticism of earlier astronomical systems'. 'The concentric spheres of Calippus and Eudoxus of Cnidos failed to demonstrate the change in the distance of celestial bodies from the Earth; although the Ptolemaic epicyclic system agreed numerically with phenomena in the heavens, it nevertheless did give rise to certain doubts. The founders of the system were able to explain the phenomena satisfactorily by introducing fictitious circles, called equants. According to them the planet moved uniformly neither in the circle in which it was carried nor in relation to the center of its epicycle. That is why the idea did not seem sufficiently adequate nor did it stand up to rational criteria'. The criticism of this aspect of Ptolemaic astronomy is central to *Commentariolus* and is used as a frame of reference to the reconstruction of astronomy:

Having become aware of these defects, I often considered whether there could perhaps be found a more reasonable arrangement of circles, from which every apparent inequality would be derived and in which everything would move uniformly about its proper center, as the rule of absolute motion requires.*

The formulation reflects the significance the author ascribed (at least with the readers of *Commentariolus* in mind) to postulates regarding compliance to the a priori principle of uniform motion when developing astronomical theories. This element has an important bearing on our reflections regarding the genesis of the Copernican theory.

* The English translation of *Commentariolus* and of *Narratio prima* taken from Edward Rosen's *Three Copernican Treatises*, Dover Publications Inc., New York, 1959.

The detailed exposition of the heliocentric theory is preceded by a list of seven principles. Acceptance of these principles was to enable Copernicus to resolve the problem to which he addressed himself 'with a far less elaborate apparatus which was more suitable to the purpose than the one that had been developed at one time'. Assumptions:

(1) There is no one center of all the celestial circles or spheres.

(2) The center of the Earth is not the center of the universe, but only of gravity and of the lunar sphere.

(3) All the spheres revolve about the Sun as their mid-point, and therefore the Sun is the center of the universe.

(4) The ratio of the Earth's distance from the Sun to the height of the firmament is so much smaller than the ratio of the Earth's radius to its distance from the Sun that the distance from the Earth to the Sun is imperceptible in comparison with the height of the firmament.

(5) Whatever motion appears in the firmament arises not from any motion of the firmament, but form the Earth's motion. The Earth together with its circumjacent elements performs a complete rotation on its fixed poles in a daily motion, while the firmament and highest heaven abide unchanged.

(6) What appear to us as motions of the Sun arise not from its motion but from the motion of the Earth and our sphere, with which we revolve about the Sun like any other planet. The Earth has, then, more than one motion.

(7) The apparent retrograde and direct motion of the planets arises not from their motion but from the Earth's. The motion of the Earth alone, therefore, suffices to explain so many apparent inequalities in the heavens.

The formulation and order of the principles may seem arbitrary at first glance. They may be easily explained by drawing an analogy to corresponding formulations that preceded Ptolemy's exposition on astronomy in the *Almagest* or, more strictly speaking, to the form in which they were presented by Peurbach in *Epitome in Ptolemaei Almagestum.* Thus for example, when Peurbach states after Ptolemy that 'in relation to the firmament the Earth is like a point', Copernicus transfers the thesis (in the fourth principle) to the orbit of the Earth and boldly announces the vast dimensions of the spheres of the stars that defy comparison with the distance between the Earth and the Sun. This is, we might add, the consequence of accepting the mobility of the Earth. The fundamental discovery of the three motions of the Earth is given by Copernicus in his *Commentariolus* in the natural order with what seems a deliberate and apparently didactic purpose for the edification of the readers. The author begins by rejecting the theory of geocentrism (second principle: 'The center of the Earth is not the center of the universe...') despite the thesis propounded by Ptolemy which holds that 'the Earth is found in the center of the firmament'.

The part of *Commentariolus* dealing with the principles in detail is not provided with 'mathematical demonstrations, reserving these for my larger work'. It begins with a description of the heliocentric system – the first in the history of science – with the planets placed in the actual order of their appearance as measured by their distance from the Sun. This enabled Copernicus to establish further only the qualitative dependence between the velocity of revolution and the distance of the planet from the Sun:

...one planet is superior to another by the velocity of its revolution, depending on whether it describes larger or smaller arcs of circles.

To explain the apparent motion of the Sun, hence the actual motion of the Earth, Copernicus employed a geometric mechanism which differed not at all from the one applied (to the Sun) in Ptolemaic astronomy. Thus the Earth's orbit is an eccentric circle in relation to the Sun. Copernicus accepts the direction of the line of the apsides and the eccentric of the orbit as constant, while changing the values given by Ptolemy. The uniform motion of the Earth along the circular orbit around the Sun is, according to the terminology of *Commentariolus*, 'the first motion of the Earth'. 'The second motion' is the rotation of the Earth around its axis 'in the course of one day... from the west to the east. By reason of this rotation the world seems to whirl in a swift motion. In this manner the Earth, together with the waters found on it and with the adjacent air, performs a rotatory motion'. By this formulation, as well as by the principles listed above, Copernicus anticipates the usual conventional argument employed against the mobility of the Earth. The argument goes as follows – the rotation of the Earth would have a cataclysmic effect upon any body that is not firmly attached to the Earth's surface.

The postulate of the third motion of the Earth (deviation, declination) put forth by Copernicus had a no less fundamental influence on the development of astronomy as the first two. Copernicus explains, among others, that the Earth's axis, inclined at an angle of 66°5 to the plane of the orbit, retained its direction in space as the planet revolved about the Sun. In this instance Copernicus resorted to the conventional concept which holds that the sphere revolves in a circular path with the celestial body steadily fixed to it. In this kind of mechanism the direction of the Earth's axis would shift in space in the course of a year. The shift is compensated by the reverse motion of the Earth's axis, the Copernican 'third motion'. As early as the second half of the 16th century there was already some understanding of the fact that no special phoronomic constructions were required for retaining the direction of the rotation of the axis and it was not this that constituted the major contribution of the Copernican idea. By establishing that 'the third motion' of the Earth's axis was completed in slightly less than the sidereal year, Copernicus was the first to explain 'the motion of the eighth sphere' in accordance with reality as the precessional motion of the Earth's axis. The irregularity of that motion, as transmitted by the tradition of medieval science, prompted Copernicus to accept the sphere of the fixed stars, invariable, immobile, as the frame of reference for the description of astronomical phenomena.

The new foundations of astronomy: the three motions of the Earth and the heliocentric planetary system exercised no obvious influence on the details of the theory of the motion of the Moon. However, as we have seen, the old theory had its defects.

Copernicus had earlier drawn attention to its inherent weaknesses and his criticism of them provided at least one point of departure for the development of the new astronomy.

As described in *Commentariolus*, the orbit of the Earth's satellite differs essentially from the Ptolemaic model. It was constructed of three circles. The largest circle – the deferent – had its center in the center of the Earth. A smaller circle, called the epicycle, carried on the circumference of the deferent, completed the revolution in the period of one month. The Moon moves along its circumference within a period of about 14 days causing a maximum deviation from the mean distance at times approximating the first and the third quarters. Copernicus thus explains the irregularities in the Moon's motion, known to Ptolemy too, without recourse to fictitious changes in distances and without introducing equants as the Ptolemaic theory had done.*

The orbits of the planets were also liberated from the construction of the equant in *Commentariolus*. The large epicycles which took the place of the Earth's orbit in geostatic astronomy vanished from the Copernican theory of the planets. As in the case of the Moon, Copernicus introduces instead the small epicycle for the planets which, together with the deferent, performed the role of the eccentric orbit. The second epicycle, with a three times shorter radius, along which the planet revolved twice in the course of one revolution about the Sun, brought the planet's orbit closer to Kepler's ellipse. This construction is mathematically equivalent to the equant and fulfills 'the leading principle of absolute motion' quoted here: the motion of the planet is the resultant of exclusively uniform circular motions.** The orbit of Mercury required additional constructions.

The characteristic of all the planet orbits in *Commentariolus* is the lack of change of their eccentricity except for the dimension of the epicycle of Mercury and for the fixed direction of the position of the lines of the apsides. Soon after Copernicus was to reject these principles taken from earlier astronomical systems. It was also to become soon evident that the detailed presentation of the new theory announced by him calls for a more complex apparatus than the one outlined in *Commentariolus*. In the conclusion of *Commentariolus*, Copernicus expressed the belief that thirty-four circles would suffice "to explain the entire structure of the universe and the entire ballet of the planets."

2

The move to Frombork about the year 1510 signified to Copernicus an increase in

* The equant was eliminated from the theory of the Moon prior to Copernicus. The 14th century Damascene astronomer Ibn-as-Shatir described the geometric construction of the Moon's orbit which is identical with the model given by Copernicus in *Commentariolus*. There are small differences in the numerical values.

** As in the case of the Moon, Copernicus had forerunners among Oriental astronomers as regards the elimination of equants. Hence, the same Ibn-as-Shatir and the earlier astronomical school of Nasir-ad-Din in Maragha (13th century). See E. S. Kennedy, V. Roberts 'The Planetary Theory of Ibn-as-Shatir', *Isis* **50** (1959), 227–235; E. S. Kennedy, 'Late Medieval Planetary Theory', *Isis* **57** (1966), 365–378.

the burden of administrative duties both at the chapter itself and in the management of the chapter estates situated in the region of Mehlsack (Pieniężno) and Olsztyn. As administrator of the chapter lands, Copernicus lived for a number of years, precisely in 1516–1519 and 1520–1521, at the Olsztyn castle. He also devoted himself at the time to broader economic questions and drew up a plan for a monetary reform. Because of the troubled frontiers with the Teutonic state, internal issues of administration assumed political weight and so there was an increasing responsibility for seemingly insignificant administrative decisions. The Polish-Teutonic conflict, as we know, entered the phase of open warfare waged in the territories of Warmia in 1519–1521. In the early months of the war, in January 1520, the chapter's administrative buildings in Frombork where Copernicus resided, which stood outside the walls surrounding the cathedral hill, were destroyed. Copernicus then moved to Olsztyn where as administrator of the chapter estates he organized the defenses of the town against the expected assault of the Grand Master in 1521. Upon the signing of the armistice, Copernicus performed the function of commissioner in Frombork.

It was in these highly unfavorable circumstances, weighed down with innumerable duties, that Copernicus began to write his major work, an extensive exposition of heliocentric astronomy, and to continue to collect the data he needed. He conducted his observations, with conventional tools which did not differ from those in use in his day. He described three of the instruments in *De revolutionibus* (*On the Revolutions*). These were: the Sun quadrant for measuring angle altitudes and a parallactic instrument (for observation of the Moon in particular); the armillary sphere enabled him to directly position the coordinates of the Sun and Moon as well as of other celestial bodies. All of these instruments were known in ancient times and the descriptions (by Copernicus) are modelled on corresponding sections in Ptolemy's *Almagest*. The solar table, parts of which have been preserved to this day, was constructed by Copernicus about the year 1517 at the Olsztyn castle and was his original idea. The sunlight reflected by the horizontal mirror of the instrument fell on the wall of the gallery. The lines drawn on it enabled Copernicus to establish the position of the Sun in relation to the celestial equator and to designate the length of time that separated the moment of observation from the equinoctial date.

The scores of observations made by Copernicus and known to us from *De revolutionibus* and from his preserved notes are not notable for their precision. Characteristic, however, is the regularity and purposefulness with which the measurements were made and utilized toward a deliberate theoretical purpose. And so a series of observations of the Sun, made in 1515 and 1516 in order to provide precise data for the theory of the apparent motion of the Sun, led Copernicus to establish that the eccentricity of the Earth's orbit was variable and that the line of the apsides was mobile. It became clear that the geometric solution presented in *Commentariolus* demanded correction. This was done in *De revolutionibus* and may be seen in the change Copernicus introduced in the Earth's orbit. The motion of the Earth is presented in *De revolutionibus* by eccentricities instead of by concentricities and epicycles.

This solution enabled a simpler presentation of the changes in the parameters of the orbit discovered by Copernicus.

It is not by accident that the observations and the theoretical work done by Copernicus on the theory of the apparent motion of the Sun coincided in time with the renewed discussion on the reformation of the calendar. The inexact reckoning of the year in the Julian calendar resulted in an error of ten days by the 16th century. The question of the reformation of the calendar was taken up by a special commission appointed at the Lateran Council of 1512–1516 at the initiative of Paul of Middleburg, who also presided over it .Paul of Middleburg was the author of two memorials on the reformation of the calendar, written in 1513 and 1516 and sent out together with a papal breve. In the second memorial he lists the names of the scientists and institutions that submitted their views on the need and the manner in which the calendar should be reformed. One of the authors mentioned was Nicolaus Copernicus. We do not know the gist of his statement. In the light of the statements contained in *Commentariolus*, that is that the equatorial year (the basic magnitude for reforming the calendar) to which 'observations made in different periods assigned a different length' is not a constant magnitude, and in the light of his manifest concentration on observations of the Sun after 1514, it might be assumed the Copernicus believed that the reformation was premature and that he postulated further research. This assumption is borne out by an excerpt of the *Letter of Dedication* written in 1542 and included in the first edition of *De revolutionibus:*

> ... when the question of the reformation of the church calendar was discussed at the Lateran Council, it was left unresolved solely because no sufficiently exact measurements were as yet available of the years and months and of the motion of the Sun and the Moon.

The postulate spurred on Copernicus in the undertaking whose aim was to work out in detail a new system of the universe:

> From that time, with the encouragment of that remarkable man, the priest Paul, bishop of Fossombrone, who directed those matters then, I bent my mind to the examination of these things.

Beginning with 1521, Copernicus resided permanently at Frombork. Up to the final years of his life, he performed a series of offices at the chapter and as general administrator was in charge of the whole diecese when the bishop's seat was unoccupied in 1523. The importance of Copernicus' activity often extended beyond the framework of purely local interests. A point in case were his activities in the area of the region's economy. In view of the depreciation of money in circulation in Prussia, Copernicus produced a study providing for remedial measures as early as 1517. As the chapter's delegate to the congress of Prussian estates, held at Grudziądz in 1522, he presented his project for monetary reform based on precursory principles. By making the money healthy, his plan aimed in effect at strengthening the union between the Prussian territories and the Polish throne. As a physician Copernicus must have enjoyed a fine reputation among his contemporaries for he was summoned to treat the bishops of Warmia and as late as 1541 left for Królewiec (Koenigsberg) at the request of the Duke Albrecht to attend an ailing courtier.

The public activity in which Copernicus was engaged clearly repudiates the term 'recluse of Frombork' by which he has been described. Yet the number of persons with whom he was on intimate terms, these principally from among the canons of Warmia, was not large. First among them was Tiedemann Giese, bishop of Chełmno (from 1538) in the final years of Copernicus' life, a friend of Copernicus for many years who shared his intellectual interests. Except for Warmia, Copernicus maintained the most enduring ties with Cracow, as may be witnessed by his contacts with Bernard Wapowski which terminated only with the death of the latter in 1535. It was at Wapowski's insistence that in 1524 Copernicus wrote a criticism *De Octava sphaera, contra Wernerum* of the dissertation *On Motion of the Eighth Sphere* (1522) of Johannes Werner. The treatise known also as *Letter to Wapowski* is, next to *De revolutionibus* and *Commentariolus*, the only astronomical text to have come down to our times. Unlike the two other works, it contains no features of the heliocentric system. In it Copernicus gave a severe appraisal of the methodological and factual errors contained in the dissertation he critized without describing his own theory of precession. He did, however, speak in general terms about his work on *De revolutionibus*:

> Finally what is my opinion concerning the motion of spheres of the fixed stars?... I have intended to set forth my views elsewhere....

At this time Copernicus had already completed the first draft of the first books of *De revolutionibus* and was writing the fourth book, devoted to the theory of precession and to the description of the apparent motion of the Sun. Revisions regarding the earlier geometric diagrams contained in *Commentariolus* were not confined in *De revolutionibus* to the modifications of the solar orbit referred to earlier, but pertained also to other parts of the work. The variability of the lines of apsides, discovered by Copernicus when he compared his own observation of the planets (beginning with his observations of Mars in 1523) with the data obtained by Ptolemy, induced him to reject the view on the fixed orientation of planetary orbits in space, expressed earlier in *Commentariolus*. Consequently, in *De revolutionibus*, when dealing with the motion of planets in latitude, he accepted for the planets orbits composed of an eccentric and an epicycle instead of the concentric circle and two epicycles. The variability of apsidal lines could be now represented simply as a uniform circular motion of the center of the eccentric.

The manuscript of *De revolutionibus* was essentially completed about the year 1530.* However, Copernicus did not yet intend to bring out his work in print. As he stated later, he was afraid of incompetent criticism:

> ...the scorn which I had reason to fear on account of the incomprehensibility and novelty of my theory, had almost persuaded me to completely abandon the work which I had begun.

* The number of books was reduced at a later time when changes were made in their arrangement. As a result the theory of precession and of the motion of the Earth were included in Book III of *De revolutionibus*.

He pointed to the examples of the Pythagoreans who did not publish their theories so as

not to expose these most beautiful things, the fruit of long and tedious research by great men, to abasement and contempt on the part of those who either resent the expenditure of honest effort for any science that does not bring them gain or... who are dull witted and have blundered among true scientists as drones among bees (*Letter of Dedication*).

As early as 1535, during Bernard Wapowski's visit to Frombork, Copernicus consented to publish the almanach which gave the positions of celestial bodies for a certain period of time computed on the basis of tables contained in *De revolutionibus*, Plans for the undertaking were frustrated by the death that very year of the man who proposed the idea. The manuscript with Copernicus' computations was lost.

Despite Copernicus' great reticence, news of his discoveries and of the work he was preparing spread beyond the borders of Poland. His discovery was discussed at the papal court in 1533. Three years later Cardinal Nicolaus Schönberg, prosecutor general of the Dominican Order who had met the astronomer during his diplomatic journey to Poland in 1518, wrote a letter to Copernicus urging him to publish his discovery and promising to disseminate it. Finally, in the spring of 1539, a young professor of Wittenberg, Georg Joachim de Porris, known under his humanistic pseudonym Rheticus,* a twenty-five-year old mathematician and astronomer, a protégé of Melanchthon, came to Frombork because of the interest that news about Copernicus had generated in Germany. He enthusiastically embraced the Copernican astronomy. A first consequence of his arrival was that Copernicus resumed work on the manuscript of *De revolutionibus*. This was made possible by the books which Rheticus had brought to Frombork. The most important of these was the Greek edition of Ptolemy (Basle 1538), much more accurate than the Latin translation of the *Almagest* (Venice 1515) which Copernicus had used until then, and *De triangulis omnimodis* (trigonometry) by Regiomontanus (Nuremberg 1533). Copernicus made certain additions to *De revolutionibus*, revised now as six instead of the original seven books, enlarging the text on trigonometry and spherical astronomy and the chapters describing the uranographic latitudes of planets. He was now decided to give his consent to the publication of the work. He was dissuaded from the opposite intent, as he wrote in the *Letter of Dedication* by

my friends... Among them the foremost was Cardinal Nicolaus Schönberg, Cardinal of Capua, distinguished in every field of learning. Next to him was Tiedemann Giese, bishop of Kulm (Chełmno), a man who loved me dearly, a close student of sacred literature, as he is of all good literature. He frequently urged upon me and, sometimes adding reproaches, demanded that I should publish and finally permit to appear the book which, hidden by me, had lain concealed not merely nine years but already four times that period. That same conduct was recommended to me by not a few other very eminent and learned men.

Having become acquainted with the Copernican theory and with the manuscript of *De revolutionibus*, Rheticus made an extensive summary of the work which was

* K. H. Burmeister wrote an extensive monograph of Georg Rheticus, called *Georg Joachim Rhetikus. Eine Bio-bibliographie*, Wiesbaden 1967–1968, Vol. 1–3.

published in Gdańsk in 1540 under the abbreviated title *On the Books of Revolution...by Nicolaus Copernicus...First Account* (*De libris revolutionum...Nicolai Copernici...narratio prima*). Of greatest interest in the *Narratio prima*, which gave an enthusiastic description of Copernicus' discoveries, are the sections in which the author recounts facts and opinions he had learned by direct contact, such as, for example, Giese's argument for the publication of *De revolutionibus*.

Most important are the motives which persuaded Copernicus to accept the motions of the Earth. They are briefly:

In the first place, the indisputable precession of the equinoxes... and the changes in the obliquity of the ecliptic persuaded my teacher (Copernicus) to assume that the motion of the earth could produce most of the appearances in the heavens, or at any rate save them satisfactorily.

Secondly, the diminution of the eccentricity of the sun is observed... in the eccentricities of the other planets.

Thirdly, the planets evidently have the centers of their deferents in the Sun, as the center of the universe .

Fourthly, my teacher saw that only on this theory could all the circles of the universe be satisfactorily made to revolve uniformly and regularly about their centers, and not about other centers – an essential property of circular motion.

The fifth argument was theological in nature. The fact that speaks in favor of accepting the 'serviceable theory of the motion of the terrestrial globe' is that it 'sufficed for the construction of a sound science of celestial phenomena...'.

Finally, the last argument invokes the harmony of the universe. '...masters of the science... fashioned their theories and devices for correcting the motion of the heavenly bodies with too little regard for the rule which reminds us that the order and motions of the heavenly spheres agree in an absolute system'.

Narratio prima was drawn up in the form of a letter to Johannes Schöner, Nuremberg astronomer, author and publisher of a series of works on astronomy and astrology. Rheticus visited Nuremberg in 1538, that is prior to his journey to Poland. At the time he had quite likely discussed with a group of friends which included, next to Schöner, Johannes Petreius and the publisher and Lutheran theologian Andreas Osiander (Hoseman), plans for the publication of the Copernican manuscript. When Rheticus left Frombork in the autumn of 1541, the manuscript of *De revolutionibus* was not yet in its final, completed form. The corrections and additions made in the manuscript in Rheticus' hand indicate that he intended to edit the manuscript before the work was printed.

Upon his return to Wittenberg, Rheticus learned that the publication of the book of Copernicus would be delayed. Responsible for the delay was, on the one hand, Melanchthon's critical attitude. As Luther a few years later, he firmly rejected the Copernican theory on the ground that it was contradictory to the tenets of the Holy Scripture*. On the other hand, in the winter semester of 1541–1542, Rheticus took upon himself the demanding duties as dean. In this period he brought out only the non-controversial *Trigonometry* (*De lateribus et triangulorum...libellus*) by Copernicus, which constituted the final part of Book I of *De revolutionibus*.

* As if in anticipation of these objections, Bishop Tiedemann Giese wrote an apologia on the work of Copernicus. The treatise has been lost.

Next to analogous works by Regiomontanus and Werner (the latter's treatise remained in manuscript until the current century), the *Trigonometry* of Copernicus represents an important contribution to the history of mathematics of that period. Written independently of the two, it yet bears a number of similarities which can be ascribed to the fact that Copernicus made use of the same ancient (Ptolemaic) and Arabic (Djabir-ibn-Aflah, known as Geber) sources. Another innovation in European science made by Copernicus, apart from the published manuscript on trigonometry, was the introduction of the function of the secant. Copernicus computed and drew up a table of this function in one of the books in his library.

With the arrival of Rheticus in Nuremberg in the spring of 1542, the Petreius press began to print *De revolutionibus*. It is not possible to refrain from expressing a certain resentment in regard to Rheticus for having submitted to Petreius an unedited manuscript which contained a series of (generally small) errors in numerical data. Of more serious consequence was his departure from Nuremberg two months later, leaving the publication of the work in the hands of Andreas Osiander.

Osiander, who had become interested in *De revolutionibus* at an earlier date, had already corresponded with Copernicus and Rheticus in 1540. At the time he suggested to Copernicus that in order to 'mollify the peripatetics and theologians whose opposition you fear', he might present the new theory as a formal hypothesis useful in conducting computations rather than as a description of the actual cosmos. Copernicus firmly rejected the suggestion upholding his position emphatically in the *Letter of Dedication* written in June 1542 and sent to the publisher. This, however, made no impression upon Osiander. Motivated by prudence or perhaps because of his convictions (the phenomenalistic approach to astronomical theories had a long history), he introduced arbitrary changes in the work. The unsigned preface that he added, entitled *To the Reader Concerning the Hypothesis of this Work*, reduced the substance of *De revolutionibus* to the level of a hypothesis:

> ...these hypotheses need not be true or even probable, if they provide a calculus consistent with the observations, that alone is sufficient... For it is quite clear that the causes of the apparent unequal motions are completely and simply unknown to this art... So far as hypotheses are concerned, let no one expect anything certain from astronomy, which cannot furnish it...

The forgery extended to the title of the book which was lengthened to read *De revolutionibus orbium coelestium.* Finally the introduction to Book I written by Copernicus was removed. It contained a praise of astronomical sciences, which 'occupy themselves with the miraculous revolutions in the universe and with the courses of stars... and with the causes of all other phenomena in heaven and finally which explain the entire system of the universe".

The printing of *De revolutionibus* was completed in March 1543, shortly before the seventy-year-old astronomer's death. Copernicus died in Frombork on May 24, 1543 after a long illness.

Nicolai Copernici Thoruniensis de revolutionibus orbium coelestium libri VI (*Of the Revolutions of the Heavenly Spheres. Six Books, by Nicolaus Copernicus of Toruń*) that is how the full title of the Nuremberg edition reads. The printed version opens with the anonymous preface by Osiander and the letter, mentioned here earlier, from Cardinal Nicolaus Schönberg to Copernicus. Large excerpts have already been quoted from the *Letter of Dedication* to Pope Paul III, which comes next in the edition. The letter speaks of the author's doubts regarding publication of the work and of the steps taken by his friends. Later in the *Letter*, Copernicus speaks of the reason why 'contrary to the received opinion of the Mathematicians and indeed contrary to the impression of the senses' he had the courage to 'conceive such a motion of the Earth'. This important passage reads as follows:

So I should like your Holiness to know that I was induced to think of a method of computing the motions of the spheres by nothing else than the knowledge that the mathematicians are inconsistent in these investigations.

For, first the mathematicians are so unsure of the movements of the Sun and Moon that they cannot even explain or observe the constant length of the seasonal year. Secondly, in determining the motions of these and of the other five planets, they do not even use the same principles and hypotheses as in their proofs of seeming revolutions and motions. So some use only concentric circles, while others eccentrics and epicycles. Yet even by these means they do not completely attain their ends. Those who have relied on concentrics, though they have proven that some different motions can be compounded therefrom, have not thereby been able fully to establish a system which agrees with the phenomena. Those again who have devised eccentric systems, though they appear to have well nigh established the seeming motions by calculations agreeable to their assumptions, have yet made many admissions which seem to violate the first principle of uniformity in motion. Nor have they been able thereby to discern or deduce the principal thing – namely the shape of the Universe and the unchangeable symmetry of its parts. With them it is as though an artist were to gather the hands, feet, head and other members for his images from divers models, each part excellently drawn, but not related to a single body, and since they in no way match each other, the result would be monster rather than man.

We discover here the most important motives that lay behind the origins of the discovery made by Copernicus: 'the designation of the equinoctial year' – calling for an accurate and strict definition of the theory of precession, listed first in order by Rheticus in *Narratio prima*; 'contradiction with the basic principles of the regularity of motion' – mentioned in *Commentariolus* and noted perhaps when Copernicus was still a student at Cracow; finally, 'the universal system and the established order of its parts' – most likely an argument developed at a later date and buttressed by the discovery that the motions of the Earth introduced an order into the description of the universe. Similar in character, it may be assumed, is the reference made by Copernicus to the precursory theories of ancients:

...I therefore took pains to read again the works of all philosophers on whom I could lay hand to seek out whether any of them had ever supposed that the motions of the spheres were other than those demanded by the mathematical schools. I found first in Cicero* that Hicetas (Nicetas) had realized that the Earth moved. Afterwards I found in Plutarch that certain others had held the like opinion. I think fit here to add Plutarch's** own words, to make them accessible to all:

The rest hold the Earth to be stationary, but Philolaus the Pythagorean says that she moves round the (central) fire on an oblique circle like the Sun and Moon. Heraclides of Pontus and Ecphantus the Pythagorean also made the Earth to move, not indeed through space but by rotating round her own center as a wheel on an axle from West to East.

Taking advantage of this I too began to think of the mobility of the Earth.

* Cicero: *Academica priora*, II, 123.

** Plutarch: *De placitis philosophorum*, III, 13.

In the final passage of the *Letter,* Copernicus attempts to anticipate theological objections pointing to contradictions between the science and the Holy Scripture:

If perhaps there are babblers who, although completely ignorant of mathematics, nevertheless take it upon themselves to pass judgment on mathematical questions and, improperly distorting some passage of Scripture to their purpose, dare to find fault with my system and censure it, I disregard them even to the extent of despising their judgment as uninformed.

He finally emphasized the serviceability of his work in the then current question of the reformation of the calendar.

The Nuremberg publishers, as we know, removed the introduction to Book I of the text proper of *De revolutionibus* most probably because of the author's expressed conviction of the value of astronomy to scientific knowledge. Thus there disappeared from the printed text the formulation testifying to the attitude of Copernicus to his predecessors in astronomy:

...I openly confess that I shall submit much here differently than my predecessors although on the basis of what they accomplished, for they were the first to blaze the path to the study of these problems.

The opening chapters of Book I describe the basic principles of astronomy given in the same order as in Ptolemy. Copernicus preceded the introduction of the principles of the motion of the Earth with the statement that in view of the immense size of the universe, it is not in disagreement with observation, 'a seeming change of place may come of movement either of object or of observer...'. Rejecting the argument which supported the theory that the earth remained motionless put forth by Aristotle and Ptolemy, Copernicus recognizes that the revolution of the Earth, together with the surrounding atmosphere, is natural (in the Aristotelian sense) and that it does not require any external cause. Referring to astronomical arguments – that the necessity to explain the irregular motions of planets speaks for recognition of the Earth as one of the planets – Copernicus modifies the Peripatetic concept of gravity by identifying it with gravitation, or attraction toward a center of each celestial body:

Now it seems to me gravity is but a natural inclination, bestowed on parts of bodies by the Creator so as to combine the parts in the form of a sphere and thus contribute to their unity and integrity. And we may believe this property present even in the Sun, Moon and Planets...

By admitting the motion of the Earth, the author continues,

finally we shall place the Sun himself at the centre of the Universe. All this is suggested by the systematic procession of events and the harmony of the whole Universe...

The law of order is one of the more important arguments against the Ptolemaic system which treats every planet independently of the rest and in which even a rational arrangement of the distances of planets from the Sun is impossible. The argument is elaborated in the famous Ninth Chapter, with its description of the heliocentric system, a system in which there is discovered

a wondrous order of the universe and an established and harmonious relation between the motion and the magnitude of spheres which can be discovered in no other way.

Most distant of all is the Sphere of the Fixed Stars, containing all things and being therefore itself immovable. It presents that to which the motion and position of all other bodies must be referred.... Next is the planet Saturn, revolving in 30 years. Next comes Jupiter, moving in a 12 year circuit; then Mars, who goes round in 2 years. The fourth place is held by the annual revolution in which the Earth is contained, together with the orbit of the Moon as an epicycle. Venus, whose period is 9 months, is in fifth place, in sixth is Mercury, who goes round in the space of 80 days.

In the middle of all sits Sun enthroned.* (*De revolutionibus* I. 10)

The descriptive portion of *De revolutionibus* ends with the 'Explanation of the Three Motions of the Earth' (Chapter 11): diurnal, annual and the motion of the inclination (obliquity) of the Earth's axis described in *Commentariolus.*

The last two chapters of the book are devoted to plane and spherical trigonometry. They formed the substance of the separate treatise brought out by Rheticus in 1542. As pointed out earlier, in writing his *Trigonometry* Copernicus referred to ancient and medieval works (*Almagest*, *Epitome*, *Elementa astronomica* by Geber).

The second book of *De revolutionibus* is primarily concerned with the subject of spherical astronomy. It is unrelated to the fundamental thesis developed by Copernicus. Spherical astronomy was included in conscious analogy to the respective chapters of *Almagest* as an element of a complete discourse on astronomy. The analogy does not extend, however, to the final chapter of Book II which catalogs the fixed stars. In this place *Almagest* presents the theory of the Sun that leads to the development of a frame of reference for the analysis of the motion of the Moon and planets and for establishing the coordinates of stars on the moving eighth sphere. Copernicus disagrees with this doctrine. He recognizes the unchanging and motionless sphere of the fixed stars as the basic frame of reference of mathematical astronomy. Coordinates of stars (the ecliptic longtitudes of stars) in the catalog of Book II are given relative to one star γ Arietis and not, as was a common practice of astronomers, to the equinoctial points whose motions, in the light of ancient and medieval observation, Copernicus considered irregular. The longitudes of stars in the catalog drawn up by Copernicus differ from the Ptolemaic tables by a constant (corresponding to the coordinate of γ Arietis). Other data were taken, at least in principle, from Ptolemy as was the arrangement of the catalog and the number of stars listed therein.

To the system created by the sphere of fixed stars, in his next book Copernicus related phenomena connected with the apparent motion of the Sun. The first part of Book III is therefore devoted to a detailed description of the Copernican theory of precession. It opens with a critical review of past efforts to explain the phenomenon of 'the motion of the eighth sphere' and its consequences: the change in the ecliptic longtitude of stars and the discrepencies between the equinoctial and the sidereal year.

For these phenomena some invented the ninth and others the tenth sphere and believed that it is due to them that all this occurs. Yet they failed to prove what they had promised. Even the eleventh sphere began to emerge, but I shall easily demonstate the superfluity of the number of spheres through the motions of the Earth.

* The English translations of the excerpts from *De revolutionibus,* Book I, taken from 'Occasional Notes', *Roy. Astron. Soc.* (1947) **2**, No. 10.

Discovery of the true cause of phenomena in the precessional motion of the Earth's axis is one of the most important features of the new astronomy of Copernicus. The careful mathematical documentation of the three motions of the Earth is based on observation data covering a period of more than 1800 years, from Timocharis to his own observations of Spica, the brightest star in the constellation of Virgo, made in 1515 and 1525. These data enabled Copernicus to establish with great accuracy the rate of the precessional motion, or to put it more precisely, its mean value, since the data used by Copernicus indicated an irregularity in that phenomenon and persuaded him to accept that periodic 'perturbations' are a real phenomenon. He explained 'perturbations' by accepting the harmonic inclination of the Earth's axis (the assumed changes in the angle of the ecliptic). He retained the principle of uniform circular motion to represent phenomena by substituting two circular motions for the harmonic linear motion. The perturbations of the Earth's axis must have, quite obviously, led to corresponding disturbances in the precession of equinoctial points. The sidereal year (the lenght of time it takes the Sun in its apparent course on the firmament to return to the same star) became the basic magnitude in the theory of the Sun (in the second part of Book III of *De revolutionibus*), superseding the equinoctial year which was measured by returns to the irregularly shifting equinoctial points. Copernicus represents the Sun's orbit by an eccentric – the center of the cirle in which the Earth moves regularly does not therefore coincide with the Sun itself. It is also possible, as Copernicus finds, to explain the Earth's motion by taking an epicycle on a circle concentric with the Sun, that means by using the construction presented earlier in *Commentariolus*. Copernicus demonstrates the equal validity of both solutions:

...there are no differences between these eventualities so long as the distance between the centers is equal to the radius of the epicycle. It is therefore not easy to decide which of the two eventualities occurs in the heavens.

Having established the motion of the lines of apsides of the Earth's orbit (the perigee and apogee of the Sun's apparent orbit), Copernicus elaborated the simple construction given above by having the center of the deferent (the eccentric) move in a small circle around the mean position. Here too he admitted an alternative solution (a concentric with two epicycles).

An when so many methods lead to the same result, I would be hard put to say which one of them is true....

The fourth book contains the theory of the motion of the Moon and the method for computing eclipses. Describing the ancient (Ptolemaic) theory of the Moon at the beginning of Book IV, Copernicus enlarges upon the objections he had raised earlier in *Commentariolus*. Naturally enough, his objections are directed at the actual (real) irregularity of motion in Ptolemy's theory (it appears both in the motion of the epicycles on the deferent and in the motion of the Moon itself). The second objection pertained to the change in the distance of the satellite. According to the theory expounded in the *Almagest*, the Moon can approach the Earth to within half of its greatest distance. The criticism of this error, as presented in *De revolutionibus*, refers to the demonstrable fact that no greater changes are apparent in the diameter of the visible Moon disk. The original theory put forth by Copernicus in *De revolutionibus* expands upon the idea described in *Commentariolus*. The concentric deferent with the Earth carries a larger epicycle along which a smaller epicycle revolves. The Moon moves on the circumference of the smaller epicycle at twice the velocity of the epicycle carried by the deferent. The epicycles cause deviations in the position of the Moon that are due to the real ellipticity of its orbit (smaller epicycle) and to the inequality of the Moon's motion, called evection, a phenomenon that had already been discovered by Ptolemy. Retaining the values of both inequalities established by Ptolemy, Copernicus drew the parameters of the Moon's orbit on the basis of observations he took from the *Almagest* and on the basis of his own observations conducted in 1500–1523. He used the observations he made in Bologna of the occultation of Aldebaran by the Moon to prove his computation of the paralax, hence also of the distances of the Moon.

As in the case of the Sun, Copernicus admits other, equally valid geometric solutions.

Keeping the proper ratios, it is possible to explain the same with the aid of eccentrics as I have done in the motion of the Sun.

Copernicus was not the first to offer an explanation of the motion of the Moon. In the 14th century, Ibn-as-Shatir (1304–1376), the astronomer of Damascus, gave an identical solution (concentrics with two epicycles) and, as he based himself on similar data, obtained similar proportions for the seperate elements of the orbit. There is no reason to believe that Copernicus borrowed the solution from him. It may be assumed that the coincidence is due to the fact that both scientists used the same frame of reference, most important being the recognition that uniform circular motion was the only admissable motion in the kinematics of celestial bodies.

The replacement of the Ptolemaic equant in the theory of planets by the small epicycle is derived from the same premise. This solution is shared by Copernicus with his Arabic forerunners: Ibn-as-Shatir and earlier, 13th century scientists of the famous astronomical center of Nasir-ad-Din in Maragha.*

The last two books of *De revolutionibus* are devoted to the planets. In Book V Copernicus examines the motion of planets in the plane of the ecliptic, separating, as did his predecessors, the analysis of the motion of planets in latitude caused by the different inclinations of planetary orbits. These problems were dealt with in the final Book VI.

The author preceded his original solution with a short description of Ptolemy's models and once again criticized the use of the equant. The chapter that follows speaks of phenomena observed in the motion of planets that were due to the motion

* E. S. Kennedy, *op. cit.* It need not be added here that the Moslem astronomers modified the theory of the motion of planets within the geocentric system.

of the Earth. The heliocentric planetary system is thus liberated from the large epicycles, used by geocentric astronomy to explain the loop inscribed in the firmament by the planets, by ascribing it to the position of the observer who moves together with the Earth.

Still left to be explained however were those deviations from the regular motion of planets which are caused by the fact that the path of planets is in all reality not an elliptic. As has been pointed out, the earlier geometric solution given by Copernicus in *Commentariolus* (the concentric with two epicycles) is replaced in *De revolutionibus* by the eccentric deferent – with only one small epicycle. By taking equal angular velocities of the planet carried round the epicycle and of the epicycle on the deferent and by selecting appropriate magnitudes and proportions of the eccentric and of the epicycle we obtain an orbit which represents the motion of the planet as accurately as the Ptolemaic orbit with the equant.

Of interest is the comparison between the old and the new models devised by Copernicus given in the fourth chapter of Book V. Stating that both are of equal value, he nevertheless chooses the eccentric-epicycle model, as it defines the eccentricities of planets in their relation to the real Sun. The parameters of planetary orbits, on the other hand, are established in reference to the mean Sun, that is to the center of the Earth's orbit. This is a carry-over from the geocentric concept of planetary phenomena. Another similar relic is the arrangement of the numerical table used in computing the visible position of planets. Copernicus did not make use of the simplification offered by the heliocentric system of coordinates but retained the Ptolemaic arrangement of tables. With these tables it was possible to obtain the geocentric position of planets at a sacrifice of accuracy.

The planets are described individually in an order ranging from the planet at the greatest to the planet at the least distance out from the Sun, hence from Saturn to Mercury. Using observations of ancients quoted in the *Almagest* as well as his own, Copernicus was able to establish the parameters of orbits, discovering by the same the mobility of perihelia and defining the rate of motion of the line of apsides.

The model of the orbit described above sufficed to explain the motion of the superior planets Saturn, Jupiter and Mars. Modifications had to be made for the inferior planets (Venus and Mercury), as the reference of the motion of planets to the mean Sun introduced a perturbation which depended on the position of the Earth. This was most clearly apparent in the motion of Venus, the planet closest to the Earth.

The orbit of Mercury called for special procedures. We might recapitulate here that, since its orbit is the most eccentric and since the planet cannot be observed except for short periods of time when it is at the greatest angular distance from the Sun, it was possible to approximate the motion of the planet only at moments close to its greatest elongation. As in the case of Venus, Copernicus accepted for Mercury an eccentric orbit with a deferent oscillating at six month periods. Within that time the planet moved along the radius vector in a harmonic motion analogous to the one introduced by Copernicus in Book III. To plot the parameters of the orbit, Copernicus used, in addition to the observations of the ancients, the observations made by Bernard Walter in 1491 and 1504.* A comparison of the two groups of observations enabled him to establish the advance of the perihelion point of Mercury, amounting to one degree in sixty-three years, 'providing that motion is regular'.' To put it more presisely, Copernicus gave two models of the orbit of Mercury in *De revolutionibus.* In the alternate solution, harmonic motion is transferred from the planet to the center of the deferent. Characteristic is Copernicus' statement preceding 'the description of another, no less trustworthy, system which may cause and explain said oscillations'.

We have here a significant example illustrating the attitude Copernicus took regarding the nature of his mathematico-astronomical theories. He fiercely defended, according to the interpretation given by Osiander, his fundamental discovery of the three motions of the Earth, the new structure of the heliocentric system, as a true picture of the world. However, as regards the datailed geometric models given in *De revolutionibus*, Copernicus takes quite a different stand. We have seen that in the case of the Sun as well as in the instance of the Moon and Mercury, he offers alternate systems, declaring both equally important in 'representing phenomena'. But, and this is most important, he does not protest the validity of the first or the second of the models described. Consequently, he must have believed that neither offered a definitive solution but that they represent an effort to find a solution within the framework of the heliocentric system.

Traces of geocentric astronomy are apparent in the theory of the motion of planets in latitude (Book VI), hence in the periodical oscillation of planets regulated by the annual motion of the Earth. An important innovation, a consequence of the acceptance of the heliocentric system, is the introduction of the inclination to the plane of the ecliptic of all the planetary orbits to replace the variable inclination of the epicycle as given by Ptolemy. In his discourse on the subject, Copernicus closely follows the *Almagest*.

Analogies with the *Almagest* have been pointed out in a number of places in *De revolutionibus*. Yet the claim formulated by certain authors that the work of Copernicus is derivative would be unjust. The objective of Copernicus was to demonstrate in the form of a complete discourse on astronomy that all phenomena may be presented and explained within the framework of the heliocentric doctrine without detriment to accuracy and that no phenomenon contradicts that doctrine. Furthermore, *De revolutionibus* is not a mechanical repetition of the structure of the *Almagest*. Wherever the new concept of the frame of reference (the heliocentric system) required it, Copernicus consciously altered the order of the discourse (Book II and III). Because of his exhaustive presentation of all the problems of astronomy, with documentation of the transition from ancient observation to theory, Copernicus as

* Bernard Walter, a burgher of Nuremberg, was the pupil and heir to the scientific legacy of Regiomontanus. Copernicus did not possess the necessary data on Mercury. As he wrote in *De revolutionibus*, V. 30 – in this respect ancient astronomers were 'aided by fairer skies where the Nile, as they say, does not exhale such fogs as our Vistula...'.

Moreover, Mercury is seen less often because of the inclination of the sphere (i.e. due to the geographic latitude of Frombork).

the author of *De revolutionibus* was universally hailed as the reformer of astronomy even before the leading scientists of the turn of the 17th century became aware of the full significance and of the consequences inherent in the fundamental discoveries made by the Frombork astronomer.

Chronicle of the Life and Work of Copernicus

1473, February 19: Born in Toruń (Thorn) Nicolaus Copernicus, son of Nicolaus, merchant of Toruń, and of Barbara nee Watzenrode, sister of Łukasz (later bishop of Warmia).

circa 1483 – the father of Nicolaus Copernicus dies.

1491, autumn: Nicolaus Copernicus signs his name in the matricula of the Academy of Cracow.

1495 – after four years as student of the faculty of liberal arts at the Academy of Cracow, Copernicus obtains, most likely with the support of Bishop Łukasz Watzenrode, the post of canon at the chapter of Frombork. Owing to the opposition on the part of some of the members of the chapter, he is not installed until 1497.

1496, autumn: Copernicus and his brother Andreas leave for Italy to pursue their academic education.

1497, January: Copernicus takes up the study of canon law at the university of jurisprudence in Bologna.

1497, March 9: Copernicus, together with the Bologna astronomer Domenico Maria Novara, observes the occultation of the star Aldebaran by the Moon. This is the earliest recorded astronomical observation made by Copernicus.

1500 – Copernicus is in Rome during the Jubilee year, most likely serving as an apprentice of law at the Roman Curia.

1500 – According to a later account of Georg Rheticus, Copernicus lectured on astronomy before a group of specialists in Rome.

1501 – Copernicus returns to Poland for a few months and obtains permission from his chapter to continue his education in Italy. This time he is to study medicine.

1501, autumn: Copernicus takes up the study of medicine at Padua.

1503, May–June: On may 31, Copernicus receives the degree of doctor of canon law at Ferrara.

1503, summer: Copernicus returns from Italy to Warmia.

1503–1510: Copernicus lives at Lidzbark at the palace of the bishop of Warmia Łukasz Watzenrode.

1504, 1505, 1507: Copernicus attends the diets of the Prussian Estates.

circa 1507: Copernicus completes the first outline of the heliocentric theory (*Commentariolus*) disseminated in manuscript copies.

1509 – publication in Cracow of Copernicus' Latin translation of the letters of Theophylactus of Simocatta, a 7th century Byzantine writer (*Epistolae morales, rurales et amatoriae*).

1510–1513: Copernicus is chancellor of the chapter of Warmia.

1512, March 29: Łukasz Watzenrode, bishop of Warmia, dies in Toruń. Shortly before his death, Copernicus takes up residence in Frombork.

1513–1516: Copernicus takes part in the discussion on the reformation of the calendar, submitting an answer to a questionnaire sent out by the Lateran Council's Calendar Commission.

circa 1515: Copernicus begins work on his major opus, *De revolutionibus*.

1516–1519: Copernicus as administrator of the chapter estates resides at Olsztyn castle.

1517, August 15: the first monetary dissertation drawn up by Copernicus.

1519–1520: Copernicus at Frombork.

1519–1521: war with the Teutonic Order.

1520, January: destruction of the chapter buildings outside Frombork castle in the Teutonic invasion. Copernicus moves to Olsztyn.

1520, August: Copernicus appointed Commissioner of Warmia.

1520, November – 1521, June: Copernicus serves as administrator of the chapter estates.

1521, August: Copernicus returns to Frombork where he takes up permanent residence.

1521–1522: Copernicus inspector of the chapter estates.

1522, March: Copernicus presents his plan for the monetary reform at the congress of the Estates of Royal Prussia.

1523, January–October: Copernicus General Administrator of the Diocese of Warmia during the vacancy on the bishop's throne.

1523–1525: Copernicus chancellor of the chapter.

1524, June 3: Copernicus sends Bernard Wapowski his criticism of Johannes Werner's *The Motion of the Eighth Sphere*. Known as *Letter to Bernard Wapowski*, the treatise written by Copernicus has come down to us in a number of copies made in the second half of the 16th century.

1528: Copernicus draws up the final version of his treatise on monetary reform (*Monnetae cudendae ratio*); he attends the conference of the Prussian Estates devoted to questions of reform held in Malbork and Elblag.

1528–1529: Copernicus chancellor of the chapter.

1529, March 12: observations of Venus, the last of Copernicus observations recorded in *De revolutionibus*.

1530–1532: Copernicus chapter inspector.

1535: at the suggestion of Bernard Wapowski, Copernicus draws up an almanach – an astronomical calendar (lost).

1536: Cardinal Nicolaus Schönberg writes to Copernicus from Rome, persuading him to publish his discoveries. Schönberg's letter is found in the first edition of *De revolutionibus*, published in 1543.

1539, end of May: Georg Joachim de Porris (Rheticus) arrives at Frombork.

1539–1540: Copernicus makes additions to the *De revolutionibus* manuscript.

1539, June: Martin Luther criticizes at Wittenberg the views of an unnamed 'new astronomer who wished to prove that the Earth moves.'

1539, September: Rheticus writes *Narratio prima*.

1540, March: *Narratio prima* by Rheticus appears in Gdańsk.

1541, August 20: Copernicus observes the eclipse of the Sun (the last known observation carried out by Copernicus).

1541, April-May: Copernicus to Królewiec (Koenigsberg) attends the sick bed of a courtier of Duke Albrecht.

1541, October: Philipp Melanchthon condemns at Wittenberg the absurd thesis of Copernicus which 'moves the Earth and halts the Sun'.

1541: Second edition of *Narratio prima* by Rheticus appears in Basle. Rheticus leaves Warmia taking with him for publication a copy of *De revolutionibus* by Copernicus.

1542, June: Copernicus writes the *Letter of Dedication* to Pope Paul III. The printing of *De revolutionibus* begun at Nuremberg.

1542: the *Trigonometry* of Copernicus (*De lateribus et triangulorum... libellus*), constituting the final portion of Book I of *De revolutionibus*, published by Rheticus at Wittenberg.

1542, December: First information about the illness of Copernicus.

1543, March: the printing of *De revolutionibus* by Copernicus is completed.

1543, May 24: Copernicus dies in Frombork.

ASTRONOMER OF THE AGE OF TRANSITION*

ALEKSANDER BIRKENMAJER †

> De motu terrae assiduo, sua opinione... praedicat tum nova illa veteribusque incognita supputatio astrorum

Wherein lies the greatness of Copernicus and why is he of such crucial importance to the history of astronomy? His greatness and importance lie in the fact that he placed in the hands of humanity the key to a problem which had baffled and perplexed mankind for thousands of years. From time immemorial, ever since Homo sapiens began to raise his eyes to the dome of heaven that spread above his head, he observed the continuous and peculiar journey of large and small points of light across that firmament. As he observed the lights more carefully, he remarked that the vast majority of the celestial bodies, notably the fixed stars, seemed to travel in a single group always moving in the same direction as if all had been fixed for eternity to a single vast sphere revolving at a uniform velocity around the Earth; other celestial bodies, on the other hand, namely the Sun, Moon and the wandering stars or planets, though they were caught up in that motion, were also endowed with certain individual motions owing to which their positions in relation to the fixed stars kept changing. Continued observation of these peculiar motions revealed that they were unusually complex; the paths of planets ran in oblique lines across the vault of heaven, inscribing a loop at times, while the rate at which they traveled along these paths was once faster and once slower. At one time the journey continued from west to east (in relation to the fixed stars) at other times from east to west and so on. The question that rose to challenge the mind of man was: How and in what manner can the maze of motions observed in the firmament be unraveled? A long register of scientists, Babylonian, Greek, Arabic and Medieval, sought to solve this puzzle. Although a series of different 'astronomical systems' arose, each more complicated than the first, yet none of them provided a satisfactory explanation of all the observed phenomena. Not until there finally appeared a thinker who offered to the astonished and incredulous world the thread of Ariadne which was to become an unerring guide through the celestial labyrinth. This ingenious thinker was Nicolaus Copernicus of Toruń.

In order to evaluate with an impartiality befitting a historian the importance of the breakthrough in astronomy affected through the agency of Copernicus, that is to say in order to pay due respect to the power of his intellect, while at the same time keeping

* Part of a paper, called 'Nicolaus Copernicus', read in Toruń at the General Assembly of the Polish Philological Society, held on June 29, 1952. The paper has not been printed in full. The French translation appeared in a volume of selected works of A. Birkenmajer, called *Études d'Histoire des sciences en Pologne*, Wrocław 1972. The title and the footnotes supplied by Aleksandra Maria Birkenmajer.

one's eyes open to the fact that even this man of great intellect was only human, hence imperfect, we must briefly review the astronomical theories, or more strictly speaking the theories of the structure of the universe, that were propounded and taught in the 15th and 16th century. They fall into two major divisions both derived from Greek thought. But while one reverted to the works of Aristotle and of his commentators, the second referred to the works of Ptolemy and his successors. Both systems agreed on certain fundamentals; two of these deserve to be mentioned specifically here. The first thesis stated that the Earth is stationary at the center of the universe, a fact of which (it was claimed) we were persuaded by the daily experience of our sense. In other words, the two systems discussed here were geocentric. The second thesis was deduced from certain a priori physical (or rather metaphysical) principles formulated by Plato. Plato, and after him all ancient and medieval philosophers and physicists, distinguished two fundamental motions: the rectilinear motion observed on Earth and its closest environment and the circular motions peculiar to celestial bodies, hence the motion of the stars and planets, among which the Sun and the Moon were also included. It was believed that, since celestial bodies possess a more perfect nature than terrestrial objects, they must be invested with motions of the most perfect order, that is with circular and uniform motions. These were the only motions, Plato declared, that were 'admissable' for celestial bodies. That statement, which we shall call 'Plato's axiom' in short, formed the second basic thesis common to the two 'systems of the universe' under discussion. 'Plato's axiom' may also be expressed as follows:

All the motions of celestial bodies observed by us, even though they appear very complex to our eyes, must be amenable to reduction to an appropriate combination of uniform motions on a circle.

On the foundations of these and similar principles, Greek thinkers built their astronomical theories bequeathed later to the Muslim culture and to Christian and medieval Europe. It is on these principles that Plato and, to a greater extent, his followers Eudoxus, Calippus and Aristotle, built an astronomical system which has been called the system of homocentric or concentric spheres. The essential components of the system are: the stationary Earth resting at the center of the universe and homocentric spheres revolving around it with a common center in the center of the Earth. The largest, or the outer sphere, revolves at a regular angular velocity around 'the axis of the universe' from east to west. This device offers adequate explanation for the motion of the fixed stars which that sphere carries round within it. To explain the motions of planets a more complex mechanism was necessary, composed of a larger number of spheres with the Earth at the center. Each of these spheres, it was said, revolved in a uniform motion around a different axis and at a different angular velocity, some of these moving in a different direction than the next sphere. By giving the axes an appropriate obliquity and the velocities an appropriate numerical value it was indeed possible to reproduce, roughly at least, the irregular motions of planets in the heaven.

Aristotle, as has been said, subscribed to this astronomical system and outlined it in his works. But in the lifetime of the Stagirite's pupils, if not earlier, Greek astronomers came to the conviction that the system did not explain all the phenomena observed in the heaven. It did not in any way explain why certain heavenly bodies found themselves closer at times and then again farther from the Earth as indicated by the detail that centric eclipses of the Sun are total at times and annular at others. It became necessary, therefore, to invent a different mechanism to explain the motions of celestial bodies, one that would stand in closer accord with observed phenomena. The new mechanism, which historians have called the 'system of eccentrics and epicycles', is more widely known as the 'Ptolemaic system' because Cladius Ptolemy, author of the *Almagest*, was the most famous propounder of that astronomical theory. The mechanism of planetary motions described in the *Almagest* is composed of two kinds of circles (or spheres): of eccentrics, i.e. circular orbits whose centers lie a short distance from the center of fixed stars, that it to say that the centers do not lie at the center of the Earth (because the Earth remains stationary at the center of the universe), and of epicycles, that is of circles whose centers move along the eccentric orbits, also called deferents. And so Mars is carried around the circumference of the epicycle and completes its revolution, known as the synodic revolution, in a given period of time. At the same time, however, the center of the epicycle is carried around the circumference of the eccentric or deferent, completing that revolution in a given time. This is called the Zodaical revolution of the planet. The motion of Mars around the Earth is complex. It is the resultant of two motions along a circle, or rather of three motions if we add the diurnal revolution of the celestial vault.

The system of eccentrics and epicycles furnished a more suitable explanation of observed celestial phenomena than the system of 'homocentric spheres'. It is no wonder then that nearly all of the astronomers who lived and worked between the ages of Ptolemy and Copernicus advocated this system, tried to perfect it and based on it their tables and calculations. Nonetheless, the system never wholly superseded or cast into oblivion the system of 'homocentric spheres'. This odd historical paradox can be explained by the tremendous authority that peripatetic philosophy and Aristotelian physics commanded for the space of many centuries. There is no need to describe the endless battles waged during those centuries by adherents of the two opposing mechanisms of motions of celestial bodies. Attention might be drawn to the fact that about the year 1200 the conflict also spread to the area of Christian Europe. Here as well two hostile camps confronted each other: the 'astronomers' who followed Ptolemy and the 'philosophers' who walked in the footsteps of Aristotle. The first as well as the second camp won and held a permanent position in the academic curriculum. The system of 'homocentric spheres' was taught when the works of Aristotle were commented while the system of 'eccentrics and epicycles' formed the basis of lectures on astronomy. Inescapably, this state of affairs led to contradictions between the two courses of lectures. But not without cause were the Middle Ages and specifically the 15th century (the period with which we are concerned) notable for scientific syncretism which was able to tolerate and which seemed to 'explain' the contradiction by taking recourse to some peculiar and often oblique methods.

But this fundamental difference between the Aristotelian and the Ptolemaic mechanism of the motions of celestial bodies was by no means the only logical stumbling block that Copernicus must have encountered in his youth. The astronomy of his day, having absorbed various doctrines of Greek, Indian, Arabic and occidental provenience, carried many equally discordant elements within its bosom. Perhaps more important still is the fact that the prince of ancient astronomers, Ptolemy himself, was not consistent in all things even with himself. As stated before, he, as Aristotle, subscribed to 'Plato's axiom', which stated that the only 'admissable' motion in heaven is the uniform motion around a circle. Furthermore, in Book III of the *Almagest* we find an entirely correct definition of that type of motion. It says in effect that in the course of this kind of motion, the radius vector from the center of the circle describes equal angles at equal times. We get a totally different picture when the author comes to the detailed description of the revolutions of planets; the center of the epicycle then moves around the circumference of the eccentric in such a manner that equal angles at equal times are described by a radius vector not from the center of the eccentric but from a different point in space. That point was later called 'the center of the equant', that is the center of a circle that equalizes motion. This same point is conceived by Ptolemy as a regulator of the planet's motion around the circumference of the epicycle.

Circular motions conceived by Ptolemy are not uniform motions arround a circle but only 'apparently uniform' motions. There is therefore a clear contradiction between 'Plato's axiom', accepted without reservation by the Alexandrian astronomer, and the practical application of that principle. Albeit Ptolemy's successors could not fail to notice this inconsistency, they nevertheless tried to make light of it and to obscure it with dialectic jugglery. The rejection by Copernicus of all sophistic reasoning, which allegedly proved that the inconsistencies in Ptolemy's system were superficial, implanted the germ which years later grew into *De revolutionibus*. Contrary to what he was taught at the universities, Copernicus had the courage to assert that inconsistencies did exist and that a thorough change if not a complete reconstruction of the system was called for. This was the first step made by Copernicus on the road which was to take him to the reformation of astronomy. That step was in essence a repudiation based on logical premises at first, on the observation and recognition of the fact that the theory of the motion of celestial bodies, accepted as truth at the time, was weighed down by the original sin of inconsistency.

This, however, was not the only rational reason which disenchanted Copernicus with the current knowledge about the structure of the universe and which aroused 'a creative dissatisfaction' rightly believed to be one of the most important motivations in scientific progress in general. Apart from the fact that he must have been irked by the wide discord between the orthodoxy of the peripatetics and the views of astronomers described earlier, it should also be added that in his time it was no longer a secret to anyone that there was no more than an approximate agreement between the *Almagest* and the heaven. The agreement was fairly good in the 2nd century A. D. during Ptolemy's life but it worsened with the passage of years; the observations conducted made this absolutely clear. Under the pressure of observations already the Arabs began to correct and supplement the work of the Alexandrian astronomer. Their example was followed by the Latinists. The 'corrections' were confined to (exiguous) changes in numerical data accepted by Ptolemy with regard to concrete values (hence the length of the equatorial year, the obliquity of the ecliptic to the equator and so on). The supplementation consisted of large and small circles added to the 'system of eccentrics and epicycles'. As a result the mechanism of the motion of the celestial bodies grew more complex and ambiguous while at the same time the inconsistencies deepened because, while individual astronomers accepted a given 'correction' or 'supplementation', others rejected them or introduced others. Theoretical astronomy was plunged into a growing state of confusion. Observations demonstrated repeatedly a disagreement between the astronomical tables and the heaven. The mind of Copernicus was therefore stimulated to look for a way out of what seemed a hopeless stalemate. Esthetic considerations played a not insignificant role in his mind as may be seen from a passage in *De revolutionibus* where, paraphrasing Horace, he says of his predecessors:

accidit eis perinde ac si quis e diversis locis manus, pedes, caput aliaque membra optime quidem, see non unius corporis comparatione depicta sumeret, nullatenus invicem sibi respondentibus, ut monstrem potius quam ex illis componeretur (It is as though an artist were to gather hands, feet, head and other members for his images from divers models, each part excellently drawn, but no related to a single body, and since they in no way match each other, the result would be monster rather than man.) (*Letter of Dedication*.)

It was these reasons, intellectual and emotional, which led the young Copernicus to the conclusion that the cosmos, a synonym for order and harmony, is not built the way the adherents of the prince of scholastic and peripatetic philosophy described it nor yet in the way the imitators of the prince of astronomy would have it. With this conclusion the period of criticism and repudiation came to an end and the period of positive search began for Copernicus. The aim was to build on the rubble of the old a new, true, consistent and harmonious system. It is quite likely that Copernicus may not have immediately realized exactly what the leading idea of this new system should be. According to his own account, he began to read avidly in the works of ancient Latin and Greek writers in order to learn whether different concepts of the structure of the universe had ever been held by anyone in the past. Indeed, in the course of his studies of the history of ancient astronomy he discovered, first in Cicero and later in Plutarch and others, vague references to certain Greek thinkers, notably the Pythagoreans, as being of the opinion that the Earth is not stationary. It was these references, discovered by Copernicus in 1498 at the earliest, that gave him the idea of accepting some motion of the Earth as one of the cornerstones of the new system.

We might pause at this point to ask: What was the historical role of the references about the Pythagoreans in the evolution of our astronomer's thoughts? Some of his biographers attach excessive weight to them and search his words for an admission that he was not the actual originator of the heliocentric system but that he only revived the cosmological concepts of Hicetas (Nicetas), Philolaus, Heraclitus, Ecphantos and primarily of Aristarchus of Samos. I shall deal with this position, at least in some of its aspects, at a later time. Right now I should like to turn my attention to those biographers who, quite on the contrary, believe that the originality of the work of Copernicus must be 'saved' by belittling what he himself says of external motivations provided by his reading of Cicero and Plutarch. To prove their point they even sought to cast doubt on the account given by Copernicus himself and assume that the motivations as described in the preface of *De revolutionibus* were placed there so as, by marshaling ex post the authority of ancient thinkers, to minimize the revolutionary nature of his work. This position should also be rejected not because the veracity of our astronomer's statements is questioned but primarily because there is absolutely no reason to question it. Whoever tries to enter the mind of Copernicus must come to the unbreakable conviction that he is a typical representative of the period of transition. He was, on the one hand, a bold innovator blazing a trail for scientific progress while, on the other hand, despite his genuis he did not completely free himself of the bonds of tradition even in those instances when that tradition, by being based on a priori premises, was clearly misleading. This shall be demonstrated by a concrete example when we touch upon the kinematic aspect of the Copernican model of the universe. I shall not mention any other examples of this nature, although I should like to state generally (and this will no doubt please every admirer of the ancients) that the father of modern astronomy was under the spell of the ancient classics. He admired the enormous achievement of Greek mathematicians and astronomers most of all. On one of the very first pages of his work, he pays gallant tribute to his principal antagonist (if I may use the term) for his 'admiranda sollertia et diligentia' (for his admirable skill and diligence) – *De revolutionibus*, I. Introduction). Elsewhere he calls him 'vir ille mathematicorum eminentissimus' (the most eminent man of mathematics) – (*De revolutionibus*, II, 14). Hipparchus appears to him 'vir mirae sagacitatis' (a man of wonderful perception) – (*De revolutionibus*, III, 1). Copernicus cherished a deep respect for the ancients not only because he was a true son of his age, the age of humanism and of the Renaissance, but also because his mind had been nurtured by the works of the ancient sages and because he believed himself to be their pupil. In view of these facts does it not seem only natural that when looking for 'certior ratio motuum machinae mundi' (a certain rate of motion in the mechanism of the universe) – (*Letter of Dedication*) – he should turn to them for counsel? There is no need to search for indirect implications in his account. It should be taken at face value. Worthy of our deepest respect therefore is the great astronomer's admission that the placita of the Pythgoreans mentioned by Cicero and Plutarch had drawn his attention to the fact that it was possible to reject the dogma stating that the earth is stationary.

This does not mean, however, that the 'few generalizations read by so many before him and perhaps considered no more than absurd curiosa'* had promptly convinced Copernicus that the Earth must indeed be endowed with certain motions. Our

* M. Ernst, Nicolaus Copernicus as astronomer in *Nicolaus Copernicus*, Lvov 1924, 1–13, p. 4.

astronomer's early attitude is clearly characterized by his next sentence which says that 'quamvis absurda opinio videbatur' (although the opinion may seem absurd), he nevertheless decided to 'ut expriretur, an posito terae aliquo motu firmiores demonstrationes... inveniri in revolutione orbium coelestium possent' (try whether by assuming some other motion of the Earth, it is possible to find stronger confirmation of it in the revolution of celestial spheres) and that only after 'multa et longa observatio' was he led to the conclusion that not only the phenomena observed in the heaven, but also the order of the universe evolve from what seem absurd premises, 'ut in nulla sui parte possit transponi aliquid reliquarum partium ac totius universitatis confusione' (that it can be altered in none of its parts without throwing into confusion the remaining parts and the whole of the universe). (*Letter of Dedication.*)

That course of action is worthy of a modern natural scientist: he accepted the possibility that the Earth moves only as a working hypothesis at first; he accepted it as truth only after he proved that it was workable.

Copernicus also rejected the Aristotelian and Ptolemaic doctrine which stated that the Earth is stationary. He nudged the Earth from the center of the Cosmos and, placing the Sun there in its stead, told the Earth and other planets to revolve about it. He did not, however, reject the second fundamental principle of scholastic astronomy, that is 'Plato's axiom' which held that the kinematics of the cosmos is composed of uniform motions around a circle. The two heliocentric systems created consecutively by Copernicus are constructed like a clock mechanism, with the small and large circles revolving in a uniform motion. In other words the deferents and epicycles, mentioned in the description of the Ptolemaic system, also found an application here. Yet the difference between Ptolemy and Copernicus lay not only in the fact that the heliocentric mechanism of motions devised by Copernicus was simpler and clearer but also, and above all, in the fact that the equants disappeared once and for all and with them the 'apparently uniform' motions which the originators of the ancient system had introduced tacitly.

I have said a moment ago that Copernicus created two systems of planetary motions successively, both heliocentric and both in complete accord with 'Plato's axiom'. The only trace of the first formulation to come down to us is a generalized outline called *Nicolai Copernici de hypothesibus motuum coelestium a se constitutis commentariolus*, written at the latest in the second decade of the 16th century. The work was not published at the time, however, because the author intended it for the eyes of a small group of his most trusted friends. I shall not stop here to discuss the differences between the first projection of the heliocentric system and its elaboration in *De revolutionibus* which, in many respects and especially as regards geometry, must indeed be recognized as a new form of that system. I should like to consider, however, the light in which we should view the objectives and effects of his decision to start almost immediately after completing *Commentariolus* to write an incomparably larger work on the same subject.

The writing and revision of *De revolutionibus* occupied over twenty years of his life. Copernicus wrote it primarily for his own use since (as we know) he hesitated for a long time before he finally agreed to have it published. But he also wrote it for others, at least for the same readers for whom he had written *Commentariolus*. He must have assumed that in spite of every precaution he may take news of his revolutionary astronomic theory would sooner or later reach the ears of his contemporaries. This is exactly what did occur. In that case then it must not have been a matter of indifference to him 'quid alii de illis iudicaturi sint' (what others will judge of it) – (*Letter of Dedication*). He must have feared most that his thesis of the motion of the Earth would be rejected out of hand by people who are incapable of comprehending the simplicity and the inherent logic of the heliocentric system, but who cling to the old views held inviolate for many centuries by tradition and based (it would seem) on the 'obvious' proof of the senses. In order to overcome these prejudices, in order to win the belief of those 'mathematicians' at least who are 'ingeniosi atque docti' (ingenious and learned) – (*Letter of Dedication*), it was necessary to demonstrate specifically that the heliocentric system does indeed provide a solution to the highly perplexing problem of the motions of celestial bodies which had baffled humanity since the dawn of history. That is why Copernicus devoted the second half of his life to intensive labor, observing, computing and writing a work which would have become the *New Almagest*, one so infinitely better than its original model!

Copernicus spent many long years in a gigantic and solitary labor to give a detailed description of the universe in what became his major masterpiece. The painstaking endeavor dedicated to that purpose was far greater than the effort that preceded the writing of *Commentariolus*, although it did not perhaps involve such great exertion of the powers of the intellect and imagination as did the original formulation of the heliocentric system. As they contemplate the second half of the life of Copernicus, some historians of astronomy are filled with a 'deep sense of regret' at the 'tragic waste of labor'*. They consider his labor 'wasted' because to the end Copernicus (as we already know) remained a faithful disciple of 'Plato's axiom' and because he accepted, without a single word of criticism, the a priori premise that it must be possible to reduce the motions of the celestial bodies to a combination of uniform motions around a circle. Just several decades after the death of Copernicus, Johannes Kepler discovered that the motions of planets cannot be made to conform to this metaphysical rule. He rejected 'Plato's axiom' and introduced his famous 'three laws' of which the first and foremost (published in 1609) stated that the orbits of planets are ellipses with the Sun in one of the foci. The intricate clock, constructed by Copernicus in nearly thirty years of concentrated effort, ran for a full two-thirds of a century.

Does that entitle us to claim that Copernicus 'wasted the labor' of a great part of his life? Does it give us the right to blame Copernicus? Most certainly not. Had Copernicus stopped at a general exposition of his heliocentric theory, had he written (and published) no more that *Commentariolus*, he would have undoubtedly suffered

* M. Ernst, *Nicolaus Copernicus* pp. 10, 13.

the same fate as the Pythagoreans who nearly 2000 years before him admitted the possibility that the Earth moved: contemporaries and later generations qualified these ideas as pure fantasy not to be believed even though they are set forth as something unique. That this same fate lay in store for Copernicus was evident from what occurred in the 16th and the 17th century. The thing Copernicus feared and which he tried to anticipate in the famous *Letter of Dedication* to Pope Paul III did in fact happen. His theories were declared absurd, worse still they were branded as heresy. Only a small handful of unprejudiced natural scientists had the courage to support them. What had convinced that handful of people? They were persuaded principally by the fact that the astronomical tables computed by Copernicus on the basis of the new theory were in closer harmony with the heaven than the tables of his predecessors.

Consequently the observations and the computations which filled the second half of his life, the labor which resulted in *De revolutionibus*, was not a waste of time. It is true that the intricate geometric constructions which Copernicus devised in order to explain the motions of celestial bodies did not stand unchallenged for a very long time. They were ruled out by the very same Kepler who had jotted down in his copy of the immortal work that it contains 'innumeras sapientiae opes'. Kepler learned astronomy and developed his ideas from these 'countless treasures of wisdom'; through the agency of the written and published *De revolutionibus*, he became the pupil of the Frombork recluse. That the pupil outdistanced the master, that he simplified the heliocentric system most admirably, is no reason why we should wring our hands at the master's tragedy. On the contrary, the fact that the master's work produced a new discovery only seventy years later provides ineluctable proof that the Herculean labor of Copernicus did not go to waste.

The intellectual history of mankind is replete with instances of epoch-making discoveries which did not immediately assume the form in which they later appear in science. It is in the nature of things that the achievements of many individuals and many generations contribute to the development of a more precise and a fuller knowledge of nature. 'Wisdom grew gradually since the beginning of the world and its growth is not finished yet', these somewhat naive, though true words, were written by Roger Bacon as far back as the 13th century. Our concept of the structure of the universe differs radically not only from the concepts of Copernicus but also from the concepts held by Kepler, Galileo, Laplace.... The great *New Almagest*, which was the goal of Copernicus when he sat down to write *De revolutionibus*, is not yet completed, nor will it ever be finished. This does not alter the fact, however, that the magnificent edifice of the astronomical lore of our time rests upon the foundations erected by the work of Nicolaus Copernicus.

NICOLAUS COPERNICUS –
FOUNDER OF THE NEW ASTRONOMY*

WŁODZIMIERZ ZONN
University of Warsaw

Omnes sua industria mathematicos iuvit

Although we are familiar with the fundamental facts concerning Copernicus and his work, history is not, nor was it ever, a closed book. This is all the more true with regard to the history of scientific discoveries, for they assume different shades and colors as the body of scientific knowledge grows. That is one of the reasons why we so frequently go back in thought to those who played a decisive role in the development of science, to the men of the Renaissance, among whom one of the brightest lights is Copernicus.

One other thing. Psychologists, sociologists and scientific historians devote a good deal of time and effort to the study of causes that make for the development of skills and character, especially in the instance of outstanding personalities, or if one wishes, in the instance of men of genius. They do so not out of a desire to pay tribute to genius. The motivation is to study and to discover the nature of the complex interactions of intellect and character with man's direct and less direct environment. They seek to discover not only facts that contribute to knowledge but also facts that are, at least in the past if not today, of utilitarian value. The reason why it is worth the trouble to examine occasionally the lives of celebrated personalities of the past is because they left a deeper and more lasting imprint on history than men of average stature.

With this thought in mind, we wish to relate facts connected with Copernicus and with the science of heaven, a branch of knowledge which he had so thoroughly and so remarkably reformed.

Astronomy is one of the oldest, if not the oldest, of all sciences. The alternate rhythm of day and night, the rhythmic changes of the seasons of the year are and were one of the most essential elements of our conscious and unconscious life (and the life of animals perhaps). In order to gain a full insight into the problem, in order to establish the duration of the rhythmic changes and to understand why night follows day and summer follows spring, it was necessary to closely observe the heaven, which serves mankind as a master clock and calendar to this very day. The demands of pure science coincided with practical demands because in every form of planning, even in the most primitive sense of the word, it was necessary to know when night

* A revised version of a paper published in the collection entitled *Mikołaj Kopernik. Studia Monograficzne* (*Nicolaus Copernicus. Monographic Studies*), published in Polish, Warsaw 1965.

would fall, how long it would last and when the spring thaw would set in. Consequently, early astronomy was the science of the computation of time.

The designation of the times of day never caused any trouble, particularly in the Mediterranean countries where our civilization took its beginning. Since the length of day and night did not alter greatly with the seasons in the Mediterranean area, there was no need to bother with them. The day was divided into twelve equal parts (regardless of how long it was light). Noon was selected as the zero point of this division because that is when the Sun stands at the highest point in its daily journey and when the shadow is the shortest. The hours were marked arbitrarily by dividing the arc along which the Sun travels into six parts to the left and six to the right of the meridian.

An estimate was then made in which of the divided segments the Sun found itself at a given moment. The length of the shadow was also used to measure time. To take an example – an individual made an appointment with someone when the shadow was twice the length of the object that cast it. The night hours were not marked, for not many people took the troble to count the hours when, to all intents and purposes, all activity was suspended.

The designation of the seasons of the year presented a graver problem. In fact this was the Achilles' heel of ancient and medieval astronomy.

The problem became extremely troublesome to the population that lived on the banks of the Nile. Since the flooding of the river exercised a decisive influence on the whole of the country's agricultural economy for the full extent of the year that followed, the inability to distinguish the seasons of the years could spell inevitable natural disaster.

We know that the seasons of the year depend, as do the times of day, on the position of the Sun in relation to the Earth's equator. Since the stars are not visible in the daytime, consequently it is not possible to establish by direct means the position of the Sun in relation to the stars. The Sun's position can be established indirectly, hence by resorting to one or another ingenious method. But at least a rough knowledge of the general rules governing the motion of the Sun in relation to the stars was required before these methods could be devised.

That must have been the way astronomy developed into a science about everything that surrounds our Earth, hence as a science of the universe.

It is highly significant that the methods employed by early astronomers remain essentially unchanged in our present day and age. The methods referred to here pertain to the construction of what is called a model but what in fact is a thought construction composed of geometric (or generally speaking mathematical) figures that represent the actual figure. In our case the actual figure is the universe. We proceed in the same manner today as the ancients did in their time and not only with reference to cosmological problems but also in the area of many other natural sciences.

Every geometric figure in this construction must reflect a given reality and never the other way around. In every model we, consciously or unconsciously, omit certain facts or phenomena that we deem of little importance to our understanding of the whole.

As a rule, reality is far richer and far more diversified from anything that the human brain can conceive.

These omissions are not always to the good. The decision of selecting what we feel is essential in a given question and of discarding what is 'unimportant', is one of the most difficult questions faced by all the theoretical sciences. No single standard or rule can be applied. We must therefore rely totally upon the intuition of the scientists and upon the confrontation of the model with later observations and experiences. I shall expand upon this subject at a later time.

In constructing a model there emerges also the matter of the conformity of its functions to the general laws of science known to man; in the case of astronomy, it is the conformity of functions to the laws of physics which govern inanimate matter. We allow for some margin of tolerance here for the simple reason that we do not, obviously enough, treat the laws of physics as absolutely infallible. The final recourse in matters of nature is after all nature itself and not the laboratory where the physicist works under fairly specific conditions which differ from those we encounter in the universe. It is true that physicists make every effort to simulate or to duplicate in their laboratories conditions that prevail in the cosmos, but simulation does not always take all factors into account. Moreover, the very formulation of the laws of physics is a human activity, hence open to possible error. In their speculations, therefore, cosmologists do not always remain in strict accord with the laws of physics. Although this does not reflect credit upon them, neither does it disqualify them out of hand.

To illustrate the point, we might take the example of the general law which had such dire effects on the Copernican system. That law stated that all motions of the celestial bodies must be uniform and circular. This example is quoted by many cosmologists who seek to defend themselves against the accusation that their concepts are not consistent with the laws of physics.

A model when first developed is 'tried on for size' to see if it fits reality, or speaking more precisely, to see if it corresponds to observations that, made at an earlier as well as at the present time, are capable of being used to demonstrate the validity of the model. Unfortunately, the past is always poorer in observations from the present and the future. Furthermore, as early observations are usually imprecise, only on rare occasions is it possible to promptly compare a model with pertinent data gathered by observation. As a rule, we must wait for the findings of new observations that are prompted or suggested by a new model. Frequently, a new model may suggest a completely new method of observation which, apart from testing the model, may make many new contributions to science. Thus a model, regardless of whether it proves right or wrong in the end, may play a positive role in the advancement of science. This applies to the ancient model to which we shall devote a good deal of attention in this account. There is no doubt whatever that the models of the universe bequeathed to posterity by ancient scientists played a positive role in the scientific investigations conducted by Copernicus as well as in the entire development of astronomy.

It is in the nature of every model to become untrue at some given point. Even though a hundred observations bear out the conditions of a given model, every new observation may furnish data that are inconsistent with and that therefore discredit the model. It is the destiny of all models to pass into oblivion sooner or later. Some die suddenly, without progeny at times, others leave a rich legacy of ideas and data which are later incorporated as constituent parts of new concepts. Finally, there are models that, with a few or more corrections and supplementations, enjoy a very long, if not an eternal, life. The science of astronomy is still too young to allow us to make any sort of prediction with regard to the current models.

In view of the repeated use of the word 'model', I owe an explanation why I have used it instead of the closely related term theory. The word theory, I feel, covers more than the term model because it includes the entire system of logical premises that justify the selection of one or another model. I avoided the word theory because its meaning is construed in fairly broad terms. The word model, on the other hand, leaves no doubt as to its meaning. Moreover, it is associated in our minds with the model in its technical sense, an association which seems both right and sound even though the technical model is not, quite obviously, identical with the word model as it is used in the cosmological sense.

But let us go back to the astronomy of ancient times. The first cosmological models were constructed with the idea of a law in mind. Treated as the fundamental law of physics by the ancients, it was proved untrue at a later time. The law states that celestial bodies move only and exclusively in uniform and circular motions.

The principle seems to have been evolved from the simplest observations of the stars. The stars do indeed move (in relation to the Earth) in circles at a uniform rate. This is a consequency of the uniform revolution of the Earth. Albeit it was discovered later that the motion of planets is not uniform or circular, the law was nevertheless retained to the great detriment of the problems of mechanics and their correct understanding, but with great benefit to all practical and formal purposes: that is the mathematical formulation of all aspects of astronomy.

The point is that ancient astronomers applied a rule which was not to be proved until the 19th century by the famous Fourier principle which states in the simplest terms that every periodic motion can be expressed by the sum of simple circular motions.

As we know, the motion of planets closely approximates a uniform motion around a circle. The same is true of the Earth which the ancients believed to be stationary. Regardless of what frame of reference is taken, the application of the Fourier principle seems all the more justified in this particular instance. We can only admire the mathematical intuition of ancient astronomers who became in this case the forerunners of the Fourier principle.

With the frame of reference situated in the center of the Earth, which in compliance with accepted terminology we might call geocentric, the motion of the Sun and of the Moon did not, at least in the beginning, raise any difficulties. The two celestial bodies revolved about the Earth at a uniform rate along the circumference of a circle; the Moon revolved around a small circle, the Sun around a large circle. That the circle around which the Sun moved had a larger radius than that of the Moon's orbit was evident, among others, from the eclipses of the Sun during which the Moon was usually in front of the solar disk. Secondly, the Moon revolved about the Earth in the course of a month, the Sun in the course of a longer period of time, completing its revolution in a year. The longer the period of revolution of a given body (hence the slower its motion in the heaven in relation to the stars), the farther it is from us. This valid reasoning was accepted by ancients by pure intuition and without any demonstrable proof.

An exception to the rule was the motion of fixed stars. They were placed on the largest of spheres which moved at the rate of one revolution per day. We know that this motion is due to be revolution of the Earth and that everything that is in the heaven must take part in this motion. But since the ancient Greeks believed the Earth to be stationary, they ascribed that motion to the spheres of the fixed star and to all the other bodies of the universe. The question that continued to baffle them was: How did the revolving motion of the sphere of the stars impart itself to all the other celestial bodies (found inside that sphere)?

Several assumptions were advanced. Some ancients, like Aristotle, suggested that the universe is filled with an invisible substance, called ether, which transmitted its motion to all the bodies. Whatever the case, the sphere of the fixed stars was believed to carry the universe within it. That is why the Greeks often thought of the motion of planets in terms of a mobile system that revolved together with the entire heaven. It is in terms of this system that we shall examine the motion of planets.

The planets, we know, revolve about the Sun along closed curves, that approximate the circumference of circles, with the Sun in the center. The planes of the orbits of all the spheres nearly merge together. We shall examine the motion of planets in projection on a common plane of motion which is the plane shown in Figure 1. The circle Z, with the radius r, represents the orbit of the Earth. The circle P, with the radius $R =$ the orbit of a planet.

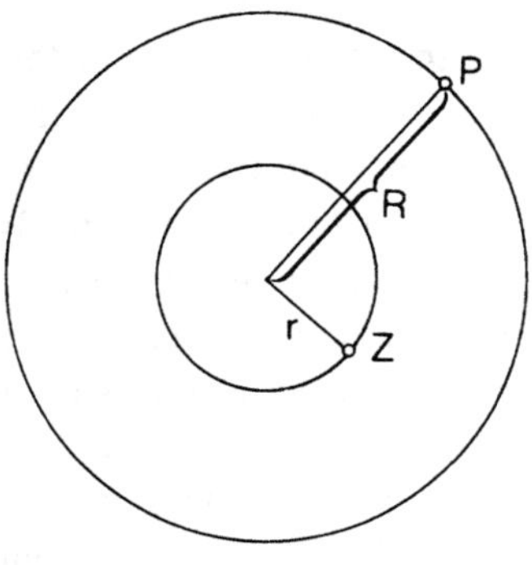

Fig. 1.

We know now that when we transfer the frame of reference from one point to another, we must add to all the motions observed in the 'old system' the motion of the new system in its relation to the old. Thus when we transfer the frame of reference

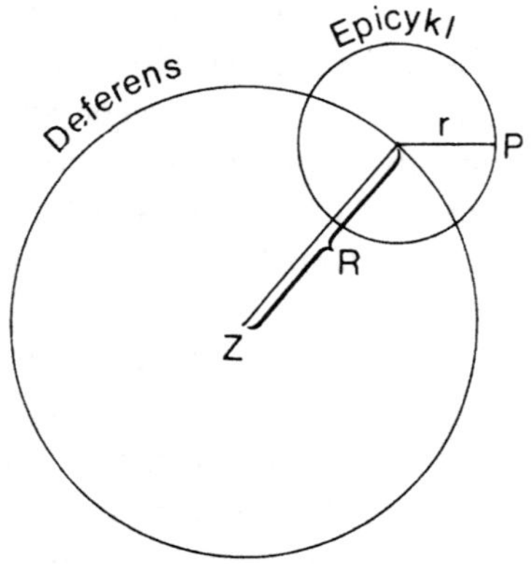

Fig. 2.

from the Sun to the Earth (returning by the same to the geocentric system of the Greeks), we must add to every motion in its relation to the Sun (a motion described in textbooks) the motion of the Earth around the Sun or around the circle with radius r (Figure 2). According to ancient terminology, the small circle with radius r is called an epicycle, the large circle (with radius R) a deferent. By selecting appropriate values for the two circles and by defining the length of time it takes the planets to move along the epicycle and of the center of the epicycle to be carried around on the circumference of the deferent, it is possible to obtain a perfect, for those days, agreement between observation and theory (if we accept the principle that all orbits are circular).

Although they had to complicate the geocentric model slightly, the Greeks managed to deal very well with the slight deviations from that principle.

Without going at any great detail into the fairly complex cosmic geometry of those distant days, I should like to give an example of a modified model which takes account of a motion that approximates the elliptic. Observed from the focus of the ellipse, it approximates the motion along the circumference of the circle when the observer is not placed at its center but out of its center or eccentrically. That is why the ancients placed some circles eccentrically in relation to the center of the Earth. They also resorted to another ingenious geometric construction. By adding (Figure 3) epicycle K_1 to circle K, they had the planet move around it in an opposite direction to the motion of the center of the epicycle carried around the circumference of circle K. Point K_1 inscribed a circle that was placed eccentrically in relation to 0.

Since the Earth's orbit is essentially an ellipse, the orbits of the planets are also slightly flattened ellipses. As data were amassed and as observation grew more accurate, the simple original geometric model had to be supplemented by an increasing number of epicycles. This was the only way the geocentric model could be 'saved'. Because of the constant improvements made on the geocentric model, by the time Copernicus arrived on the scene, the universe was described by about 80 large and small circles.

At the time when the geocentric model was originally developed, it was doubtlessly a positive and creative step because it was serviceable in the calculation of time and in

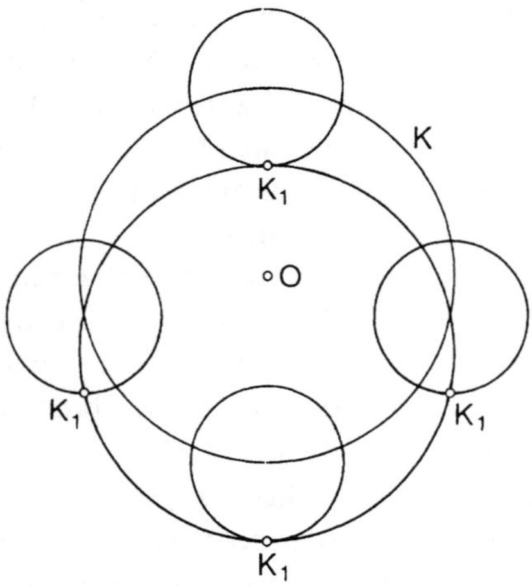

Fig. 3.

the forecasts of the seasons of the year. Moreover, as every scientific theory, it provided incentive for new observations and for the revision of ancient data. Furthermore, it was based, as we have already pointed out, on a very clever, if not 'ingenious', mathematical concept.

Although the geocentric model might be criticized for having 'falsely' placed the Earth at the center of the universe, the 'fallacy' exists only in terms of Newton's mechanics. In the light of contemporary physics, there was no fallacy or error. Nature distinguishes no single frame of reference, nor is there a single point at which the system is perfect or 'true'. The choice of a given frame of reference is therefore purely a human matter. And since we are motivated by the human factor, the first and wholly sensible idea was to place the point of reference in the eye of the observer. We follow the same principle today when we lecture to beginners on elements of spheric astronomy. That the geocentric system proved inconvenient at a certain point of time is something else again. It was inconvenient because firstly, the planets move along ellipses and not in circles, and the description of elliptic motion within a frame of reference that also moves along an ellipse is a singularly complicated undertaking from the mathematical point of view; secondly, the geocentric system would have proved terribly troublesome for those who later developed the principles of the dynamics of the universe.

Yet, the geocentric theory could not have been said to be fallacious from this point of view either. In modern dynamics, as we know today, no single system is distinguished or favored by nature. We may take the Earth as the starting point of the system today as well and, contrary to Newton's principles which, incidentally, do not describe facts accurately, construct the dynamics within the geocentric system. We would not be in error, we would only encounter more mathematical problems and difficulties than if we selected the Sun as the point of reference.

In the light of Newton's mechanics, the geocentric system was a transgression because it had not been distinguished by nature; because it was not an inertial system. The Sun, on the other hand, came closer to the Newtonian ideal where the laws of classical mechanics conceived by Newton were in force. Unfortunately, or perhaps fortunately, the concept of the inertial system has become an anachronism (from the viewpoint of natural science alone for it has otherwise remained a convenient methodological concept with a fairly limited applicability).

It is not the purpose of this account on the choice of a frame of reference to offer an apologia of the geocentric system or to rehabilitate it, for there is no need to do that. The purpose is to replace the 18th century point of view which prevailed in the past and which, unfortunately, dominates the minds of the broad public to this day, with a thoroughly modern approach to the subject.

The geocentric model obstructed further progress in perfecting the theory of the universe not only because of the inconvenience the frame of reference represented from the methodological point of view, but also because the model assumed that all motions were uniform and circular. Every fresh discovery of a planet's deviation from the course established by the geocentric theory made it necessary 'to save 'the discovery by adding a new circle to the already existing mechanism of the world.

The heaven may be compared to a clock. The imperfections of a clock's mechanism are not apparent right after it is set in motion. The faults and shortcomings are discovered only days and weeks later. The geocentric model worked very well right after it was put in motion. Later, however, in medieval times and at the beginning of the modern era, that mechanism of the world grew obsolete.

In the 11th century, Djabir (Geber) of Seville spoke out critically against the numerical data that described the geocentric model. It's not surprising that his criticism failed to evoke any greater changes in astronomy.

The famous Averroes of Cordova, called the Arabic Aristotle, took even stronger issue, yet his discussion of the Ptolemaic and the Aristotelian geocentric models did not lead toward a reformation although it did stir up much animation in astronomy.

In the 13th century Alfonso X (died 1284) king of Léon and Castile, convoked a college, composed of sixty of the leading Arabic and Jewish astronomers, and instructed them to make a fresh computation using all the astronomical data compiled so far. The result were the famous Alfonsine tables which gave numerous data without offering an explanation why certain old values were replaced. The astronomers, it would seem, were guided by the desire to acquit themselves in an exemplary fashion of the task placed upon them by the wealthy monarch rather than by a wish to study reality.

Although science did not flourish in the Arabic world at the time, yet it did not languish as it did in the Christian world. Here nearly all attention was concentrated on astrology. Thanks to it, exact astronomical observations were not entirely discontinued, but that was about the whole extent of its contribution.

There is no doubt that the Copernican reform, borne, to put it bombastically, on the turbulant waves of the Renaissance, was the turning point, not only in the history of astronomy but also in the history of all the sciences. It might not be easy to understand at first how an idea which changed the ancient theory of the universe quite thoroughly could possibly arise in a period when admiration of the ancient sciences was taking hold.

The Renaissance marked above all a return to the mood and the climate that prevailed in ancient times; a return to a view of a world free of the prejudices and restrictions that science and knowledge accumulated over the long years of the Middle Ages. But a return to that mood was not tantamount to a blind imitation of ancient science. On the contrary, certain differences among astronomers (I have in mind Ptolemy's dispute with Aristotle) prepared the ground for fresh efforts at revising the theory of the universe. We must remember, however, that an idea or a concept can be rebutted by a well-grounded argument alone; that argument could not be advanced until the age of the Renaissance.

Two German scientists were the precursors of Copernicus. They were: Georg Peurbach and his pupil Johann of Koenigsberg, known as Regiomontanus. The two lived and worked in the 15th century, a period of burgeoning humanism. Through their efforts, the Ptolemaic system, known only in its dogmatic and fossilized form, was transformed into a viable scientific theory. Since the two astronomers were not satisfied with the old translation of Ptolemy's work, they searched out the original and, having found it, produced a new translation. Simultaneously, they provided an interpretation of Ptolemy's theory which introduced much that was fresh and original.

In spite of their efforts, however, the geocentric concept of the universe remained as antiquated as ever. Confusion still reigned supreme in astronomy. The situation deteriorated further as commerce with overseas countries and the swift advance in the art of navigation placed urgent demands upon the astronomers. Geocentric astronomy was not equipped to cope with them just as it wasn't capable of dealing with the demands of astrologers.

In the lifetime of Copernicus, Christopher Columbus discovered America, sailing three poorly fitted ships. The New World, which just so recently was a country of mystery and legend, wrapped in the noxious clouds of eternal fog and peopled by one-eyed Cyclopes, suddenly became a concrete geographic concept. And so the Earth, once the habitat of mystical and legendary creatures, now became a very tangible reality. But the Earth was not yet recognized as a planet. Its central position in the universe belied such an idea.

The new universe was still undiscovered, still governed by laws that were an odd mixture of ancient astronomical assumptions and medieval dogma and prejudice, a mixture of truth and falsehood.

The discovery of a new universe called for greater (and for a different kind of) insight and knowledge than Columbus had. These rare attributes were happily united in a humble canon of Frombork who called himself Nicolaus Copernicus of Toruń.

We all have a fair idea of the contribution made by Copernicus to astronomy. We know that 'he stopped the Sun and moved the Earth', or to put it in terms of modern physics, that he tried to describe the whole of cosmology by using a different system of reference, that is specifically by selecting the Sun as its beginning. This step enabled him to clear up many puzzling problems and obscurities and to substantially simplify the mechanism of the universe. By doing so he had to automatically discard everything that from the geometric point of view constituted the quintessence of the geocentric system and that was its greatest virtue: namely, the application of the Fourier principle to astronomy. Being a remarkable mathematician, Copernicus fully appreciated that loss.

Although Copernicus placed truth and the sense of reality before the virtues of ancient science, he did not break with ancient science completely but clung to the false belief that the motions of celestial bodies must be uniform and circular and gave expression to this belief in his system, crippling it to some extent. In every instance that the elliptic orbit became discernible (in the sense of exact astronomical observation), Copernicus was forced to add epicycles or to complicate his system by moving the center of the orbits from the Sun to some point nearby. As a result, the Copernican system was not as simple as we might imagine it to be at times. It contained about forty circles of varying dimensions. Long years of tedious and unrewarding labor on the part of Johannes Kepler were necessary before the last vestiges of the old theory could be swept away from the heliocentric system. Studying the motion of Mars, Kepler based himself on data amassed in the course of long years of observation by his master Tycho Brahe.

The new feature of the Copernican system was that the Earth lost its central position and became one of the planets. This dethronement of the Earth carried with it enormous philosophical consequences.

Man, who had been accorded a privileged position by ancient astronomers, was now assigned to a totally different role. Man, the Great Solitary Being of the Universe, was one no longer. Living on one of many planets, man could no longer claim to play the major role in the universe. Common sense suggested the idea to him that creatures like or similar to him may live on the other planets. Because everything is governed by the same natural laws, the concept that there are things of greater or lesser significance in nature loses all meaning. The law that governs the motions of the Earth also governs the motion of a rock thrown at a window by a hoodlum. Although the motion of the Earth is not comparable to the motion of the rock that breaks the window, that circumstance has no bearing on the present case.

It is hard to say if by moving the Earth from its foundation, Copernicus had foreseen to what extent that step would affect man's philosophical outlook. The fact that this subject is not mentioned in his writing cannot be taken as proof that the astronomer was unaware of these matters. Copernicus was an unusually reticent man, he carefully weighed every statement he made, especially in the area of philosophy. It was not a field in which he was competent or active. Consequently, his attitude on this question will remain one of the unsolved riddles of history.

The theory formulated by the Frombork canon placed the Sun in a very important role, making it the central body of our planetary system. Fairly soon afterwards it became clear that the central position of the Sun in our system did not mean that it occupied this privileged position in the universe. The celebrated Giordano Bruno conjectured as much only several decades after the death of Copernicus. The one and only reason why the Sun is the central body or our planetary system is because it possesses a larger mass than do the other planets.

I am sure that it will never be discovered exactly how geniuses arrive at their astounding discoveries. This is particularly true of Copernicus, who, as has already been pointed out, was an unusually reticent and cautious man about his private thoughts and emotions. That is why a remark contained in *De revolutionibus* deserves our special attention in this regard. It is the passage where Copernicus explains why he believes that his order of the universe is more valid than the geocentric system. He says:

So in the course of their exposition, which the mathematicians call their system, we find that they have either omitted some indispensable detail oı introduced something foreign or wholly irrelevant. This would of a surety not have been so had they followed fixed principles; for if their hypotheses were not misleading, all inferences based thereon might be surely verified.

Georg Joachim Rheticus, a pupil of Copernicus, seems to follow his master's line of reasoning when he says:

Aristotle said that a higher truth is one whose consequences are in turn true. In accordance with this, my Master employed hypotheses which were capable of confirming the validity of observations of past ages and which, we expect, will provide the foundation for making correct predictions of all astronomical phenomena in the future....

Neither consistence with philosophical views nor consistence with common sense are the criteria for the validity of scientific theories. The main criteria for the validity of scientific theories are consistence with observation and the ability to make valid predictions.

The heliocentric theory did something more than just satisfy these criteria; it set astronomy, and indirectly other sciences as well, on the path toward the right goals by liberating it from what were totally alien philosophical and religious shackles. The theory, therefore, fully deserves to go by the name of Aristotle's 'highest truth'. Copernicus never used that term himself not because he was unaware of the validity or of the significance of his ideas, but simply because he was too modest.

But let us return to matters more closely connected with astronomy. There is in the work of Copernicus an intuitive perception of the principle which denies the idea of the absolute in physics and astronomy and which to this day remains a highly important factor of science. We have in mind the principle of relativity of motion formulated by Galileo a hundred years after the death of Copernicus. It was not until the 20th century, however, that scientists finally took up the problem of relativity in other fields of physics. As a matter of fact the 'relativization' of the whole of physics has not yet been completed.

As has already been pointed out, Copernicus was the first in the history of science to shift the frame of reference from one point to another and to foresee all the kine-

Fig. 4.

matic consequences of this step. For the benefit of those who may not be extremely well versed in physics, we might illustrate the principle by taking the very simple example of a passenger in a moving railway carriage with transparent walls throwing a ball in the air. The passenger sees the motion as rectilinear while the station master whom the train passes sees a different motion. To him the motion is curvilinear, as seen in Figure 4. A person observing the ball from the Moon surface will give us yet another description of its motion. The important thing here is that each of the descriptions is correct and valid.

There are persons who might point out that the passenger's description is best because in his system the motion can be described in the simplest terms, that is to say that in his view the motion is rectilinear while in a every other system the ball moves in a curved line. Indeed, the first description is the best, but only from the specific human point of view. Nature, however, is not created in the likeness of the ideas of man and that's why all descriptions of motion are of equal importance.

This fact would indicate that motion is a relative concept, that it all depends on how an observer behaves in relation to the moving object.

Although Copernicus did not discover the principle of relativity of motion in the present sense of the word, he was no doubt its forerunner. Furthermore, he was the first in the history of science to perform the intellectual feat of transferring a system of coordinates from one body to another.

A seeming change of place may come of movement either of object oı of observer or again of unequal movement of the two (for between equal and parallel motions no movement is perceptible)

Copernicus wrote in *De revolutionibus* (I.5). Later he quotes a line from a poem by Virgil::

We sail forth from the harbour, and lands and cities retire.

He goes on to say:

As the ship floats along the calm, all external things seem to have the motion that is really that of the ship, while those within the ship feel that they and all its contents are at rest ... (*De revolutionibus*, I.8).

'Now it is from the Earth that the rotation of the Heavens is seen. If then some motion of Earth be assumed it will be reproduced in external bodies, which will seem to move in the opposite direction.

Consider first the diurnal rotation. By it the whole Universe, save Earth alone and its contents, appears to move very swiftly, Yet grant the Earth revolves from West to East, and you will find, if you ponder it, that my conclusion is right.' (*De revolutionibus*, I. 5).

Then again we come to a quotation, this time from Cicero writing about ancient theories (how fond Copernicus was of quoting the ancients):

All of them made the Earth rotate in the midst of the Universe, believing that the Stars set owing to the Earth coming in the way, and rise again when it passed on.

Copernicus goes on to say from himself that every motion observed in the vault of heaven does not come from it but from the motion of the Earth. The Earth with its nearest elements (he must have had the Earth's atmosphere in mind – W.Z.) revolved in its unalterable poles in the course of the day; the vault, on the other hand, together with the highest sphere, remained stationary.

'Would it not then seem absurd to ascribe motion to that which contains and locates, and not rather to that contained and located, namely the Earth?'

If it is the Earth that rotates about an axis and not the heaven that revolves, then the diurnal motion of the stars we observe is a relative motion which is the result of the Earth's rotation. In that case the motion of the stars must remain the same no matter how we distribute them, whether we arrange them on the surface of a sphere or whether we disperse them arbitrarily in space. In both cases the motion observed from the Earth will be the same: one revolution per day. In the light of this simple line of reasoning the sphere of the fixed stars becomes pure fiction.

Although we do not know whether Copernicus followed this line of reasoning or not, the fact is that the sphere of the fixed stars, yet another legacy of ancient astronomy, was retained by him as a feature of his system. We have here one more illustration of the hypnotic hold that his master, Ptolemy, had on Copernicus even though he repudiated the master's system. There are certain indications that Copernicus came very close to disavowing the concept of the sphere of the fixed stars. In *De revolutionibus* (I.8) he quotes a hackneyed Aristotelian statement which assumes significance in the case in which he cites it:

...the infinite can neither be traversed nor moved....

But lacking direct evidence and being cautious by nature, Copernicus adds:

Let us then leave to Physicists the question whether the Universe be finite or no....

Whatever his attitude, the fact remains that the Copernican theory led directly to the elimination of the sphere of the fixed stars. Indeed, it soon vanished from astronomical works. Thus then due to Copernicus and his work, the human race found itself in the vast and limitless universe of modern astronomy.

Having accepted the principle of relativity of motion in reference to the Earth and to the fixed stars, it became obvious that the principle should be invoked to explain the motion of planets; we proceed as we had in Figures 1 and 2, only that this time we reverse the order from the geocentric to the heliocentric system. Copernicus arrived at the conclusion that, in addition to the diurnal motion, the Earth must have an annual motion around an orbit which Copernicus conceived as a circle and which in fact is almost a circle (a conviction that his contemporaries must have thought more audacious than the concept that the Earth rotated around its axis). The Earth and all the planets revolve about the Sun along the circumference of circles with different radii and at different time cycles.

Formulated in this manner, the statement explains many obscure and mystifying problems. In point of illustration, we might take the example of the two planets Mercury and Venus which always remain near the Sun while other planets move away to any arbitrary angular distance. We might also take the example of the odd and baffling similarity in the epicyclic motion of the three planets Mars, Jupiter and Saturn: the radii joining these planets with the center of their epicycles are parallel to each other at all times and parallel to the direction Earth–Sun. Finally, we see that the Moon in its quadratures has the same angular diameter as in its full and crescent phases, a fact which contradicted the geocentric theory.

The complicated machinery of epicycles was no longer needed in the heliocentric system. For if the Earth moves along the circumference of a circle in the course of a year, then that motion must indeed 'reflect' upon the motion of all the bodies observed in the heaven, hence also upon the motion of planets. To the progressive motion of the planets around the Sun, one must add the motion of the observer – the Earth. In this manner the motion of planets observed in the heaven is the sum of two motions that give the loop of retrogression, a motion long known to astronomers. As a matter of fact ancient astronomers devised their system of epicycles in order to explain this phenomenon.

We have discovered in this system a wondrous order of the universe and an established and harmonious relation between the motion and the magnitude of spheres which can be discovered in no other way. (*De revolutionibus*, I. 10.)

The planets move around the Sun in orbits which Copernicus conceived as circles and which in fact are nearly circles.

The planet nearest the Sun and therefore one that moves the fastest is Mercury. The planet received its name from the ancients perhaps because its motion was the swiftest. The second planet out from the Sun, shining with a lovely bright white light, is Venus.

Both planets as observed from the Earth (Z) never get very far away from the Sun (S) (see Figure 5). For Venus (P) the greatest angle α distance from the Sun (S) is about 45°, for Mercury – 15°. The planets are visible only in the evening and in the morning. They are never seen at midnight because the Sun is then far behind the horizon. The planets have been called evening or morning stars although the name is commonly applied only to Venus. Mercury, because of its nearness to the Sun, is rarely visible with the naked eye.

(Legend has it that even Copernicus never had occasion to observe Mercury. This seems highly unlikely, however, considering that he lived many years in Italy where the weather is far better than on the Gulf of the Vistula, which he mentions in his work as the place where he made observations in the latter part of his life.)

The planets farther out from the Sun: Mars, Jupiter and Saturn, whose orbits lie

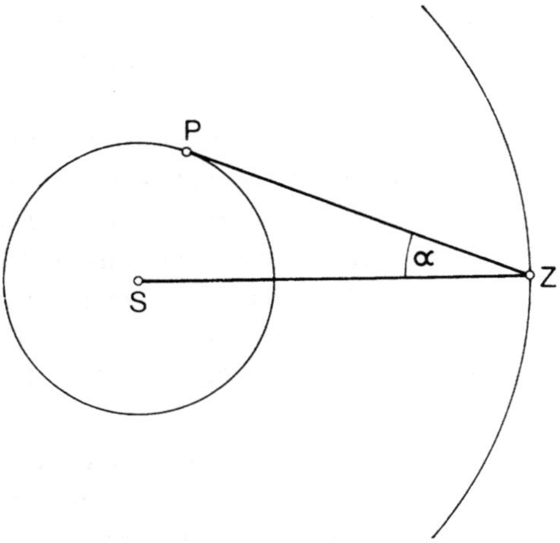

Fig. 5.

outside the Earth's orbit, behave differently. In the light of the Copernican theory, the division of the planets into two groups seems logical and reasonable because that is the way the Earth's orbit divides them.

The outer planets, Mars, Jupiter and Saturn, describe loops in the heaven, always moving at a slower rate than the Sun. The minor planets, on the other hand, once overtake and once drop behind them. The phenomenon was at odds with the geocentric theory.

Only the motion of the Moon did not undergo any serious changes. The Moon continued to revolve around the Earth 'as if in a small epicycle', as Copernicus put it. But, together with the Earth, it revolved around the Sun. This was a new idea. The Earth and the Moon thus constitute a miniature solar system, a sort of sub-system. This, in the most general terms, is the idea of the Copernican heliocentric structure of the universe enclosed in the sphere of the fixed stars. Did the idea have any antecedents? Or should priority in the discovery of a new universe be awarded to Copernicus?

Archimedes and Plutarch did indeed mention the fact that the famous Greek astronomer Aristarchus of Samos had stated that it was allegedly not the Earth but the Sun which was the center of the universe.

These reports are wholly trustworthy, most particularly the reference made by Archimedes, a contemporary of Aristarchus who in a sense was his younger colleague. Archimedes wrote that Aristarchus of Samos

> assumed that the fixed stars and the Sun were stationary while the Earth revolved about the Sun in a circle.

Except for these references there is not another trace of the heliocentric idea to be found anywhere, not even in the works of Aristarchus of Samos who had left behind him an extensive astronomical work, called *On the Dimensions and Distances of the the Sun and the Moon.* This work contains not a single word about the heliocentric idea.

The above would indicate that Copernicus could learn no more than was contained in the brief reference. It is also highly probable that he never read the remark. Had he read it, he would not have failed to mention it in *De revolutionibus* when he presented arguments in defense of his theory against eventual objections in the future. Copernicus mentioned all (not numerous) ancient astronomers who expressed certain doubts in their works regarding the central position of the Earth in the universe. They were: Philolaus, Heraclitus, Hicetas, Ecphantos. There is no mention made of Aristarchus.

But not that is the point at issue. The reading of the reference about Aristarchus of of Samos could not have changed the situation greatly, for one reading of a sentence does not necessarily transform an ingenious-minded man into a genius. Only those who have no connection with science and learning may imagine that great ideas are 'inspired' by chance phrases read or heard somewhere. The inception of a new idea is an incredibly complex and intangible process, to those who have conceived it as well. One thing is certain, however, and that is that whatever comes from the outside must fall on well prepared ground in the mind of the given scientist, otherwise every person who had seen an apple fall could have discovered Newton's theory of gravitation.

In the case of Copernicus, the ground was prepared by ancient science, irrespective of the doctrine that the Greek and Egyptian scientists had propounded. As has been pointed out earlier, Copernicus was deeply affected by ancient science.

The lack of success of the idea put forth by Aristarchus of Samos in the age in which he lived and worked is a totally different consideration. Although we do not know what arguments that remarkable scientist used to support his assumption, we might be sure that they were not trivial. Nonetheless, the arguments he might have advanced failed to convince ancient astronomers for reasons we can easily guess.

Ancient astronomy was very closely allied with mathematics. Cosmological models were judged principally by mathematical standards. In its time, the geocentric theory, with its epicycles, was so perfect in the formal sense and so lucid in its ideological content that from the mathematical point of view if rejected it would have to be replaced with something less perfect. And Artistarchus of Samos could not suggest anything of the sort.

Nor could Copernicus suggest anything, although a different mood and a different ideology prevailed in his time. The chief motto and the principal criterion of the validity of a theory was agreement of theory and observation. That criterion later led Kepler to renounce the so convenient (in the mathematical sense) circular orbits. The criterion compelled Galileo to conduct scores of experiments and scores of observations of the stars and Sun through a telescope (observations which might have caused his partial loss of sight). In this atmosphere, the ideas of the Polish astronomer, supported by vast quantities of data, marked a truly important turning point (which fact does not belittle the ingenious and perceptive insights of Aristarchus of Samos).

Acting with fairly logical consistence upon the principle of relativity of motion (which Copernicus never formulated), Copernicus gave yet another, equally revolutionary, formulation which is not widely known save to a small handful of astronomers. That is his interpretation of precession, a phenomenon already known to Hipparchus but which was explained as the motion of the sphere of the fixed stars. Later, it was thought that the motion ought to be ascribed to more distant spheres and that is how the ninth and even the tenth sphere were introduced. Quite justifiably, Copernicus ascribed that motion to the Earth, committing an incidental error which was easily discovered by his successors Galileo and Kepler. What interests us, however is the idea of transferring the motion of the sphere of the fixed stars to the motion of the Earth, providing yet another argument in favor of eliminating the sphere of the fixed stars which, as we have noted, soon vanished from astronomical considerations.

There is one other matter connected with this, a matter of immense importance not only to astronomy alone. The ancients, who treated the sphere of the fixed stars as a moving object (owing to the precessional effect), could not relate their observations to it. In other words, they could not take the sphere of the fixed stars as a system of reference. This circumstance further fortified them in the conviction that the Earth is the only system of reference. The equinoctial point and not a star was chosen as the direction of one of the axes. The position of the equinoctial point is connected, as we know, with the position of the Earth and not with the position of the stars.

Aware of the motion of the equinoctial point (owing to precession), Copernicus discovered the proper system of coordinates for astronomy. The system is based on the stars, or to put it into more modern terms, on the mean position of many stars. Not the position of the stars should be related to the point of the equinox, that is to the point which changes position in the course of time, but on the contrary, the position of the point of the equinox should be related to the sphere of stars.

Following the example set by Copernicus, astronomers began to make use of the system of reference which was later called Galilean and which also plays an important part in physics. It is a 'universal' system in the purest sense of the term, for it completely sets apart the problems of our planetary system from considerations of the entire universe. It is a system which we use to this day (although we are trying to introduce certain modifications by replacing stars with distant galaxies).

According to the heliocentric theory, the Earth revolves around the Sun once a year in an orbit which Copernicus conceived as a cricle and which in fact deviates but slightly from the circle. The Earth's motion as foreseen by the statement made by that genius Aristotle, and by Copernicus, should be 'reflected' in the motion of all the celestial bodies, hence also in the motion of the stars. In the course of a year, every star should inscribe a larger or a smaller ellipse in the heaven, depending on its distance from the Sun.

The fact that ancient observers did not discover these motions was the reason why Aristotle was deflected from consideration of any heliocentric hypotheses. Copernicus gives a different interpretation to this failure which wholly accords with truth:

That there are no such phenomena for the fixed stars proves their immeasurable distance, compared to which even the size of the Earth's orbit is negligible and the paralactic effect unnoticeable. For every visible object has a certain distance beyond which it can no more be seen.

Today we would simply state that with small angular measures every object becomes a point. The twinkling light of stars leads us to believe that a vast distance lies between the highest planet, that is Saturn, and the sphere of the fixed stars. The twinkling light represents the clearest single distinctive trait between the stars and the planets. This trait also indicates, as it should, the greatest distinction between moving and stationary bodies. "So great is the divine work of the Great and Noble Creator!" (*De revolutionibus*, I.10).

We shall pause here to offer a short explanation of the passage that ends with a magniloquence which seems hardly justified by the substance of the paragraph preceding it. Whether an object found beyond the Earth's atmosphere twinkles or emits a steady glow has no direct relation to its distance from us. The quality of light depends exclusively on the object's angular measures. If Mars or Venus had very small linear, hence also angular, dimensions, their image would perhaps twinkle as do the stars. The angular diameters of larger planets are of the order of one-tenth of a second. Although this angle is not enough to make the planets visible as a disk to the naked eye, yet it is sufficient to reduce the twinkling effect of the light they emit. The angular diameters of stars are much smaller, less than a hundredth part of a second, which fact produces the twinkling effect, most noticeable when turbulence is present in the Earth's atmosphere.

But let us return to the subject of the annual motion of the stars in the heaven. The motion is due to the Earth's revolution around the Sun, that is to what the astronomers have called the parallax effect.

As soon as astronomers became acquainted with the works of Copernicus, they applied themselves to work with the hope of discovering the parallaxes which would have provided the most convincing and obvious confirmation for the validity of the Copernican theory. However, the first decades did not bring any results. Two possible explanations may be given. The failure either repudiated the heliocentric theory, a point of view embraced by Tycho Brahe who created his own concept of the universe which constituted a compromise between the Copernican theory and the geocentric concept, or we might accept Copernicus' view that the stars lie too far from us, much farther then the ancients assumed. The second explanation found credence with the majority of astronomers who came after Copernicus. It was for this reason as well that they began to pay greater attention to accuracy in astronomical observations. Yet, every effort ended with failure proving that the distance to the stars was greater than previously assumed. Thus the universe continued to grow in magnitude with each unsuccessful attempt to measure the parallax of the stars.

New and astonishing discoveries were made, however, discoveries that were of greater importance to astronomers than the original object of their observations. Bradley discovered the aberration of light which provided telling evidence of the finite velocity of the propagation of light in space and which offered better confirmation for the validity of the heliocentric theory than would the stellar parallax. Observation of the aberration of stellar light enabled the measurement of the velocity of the Earth's motion, in its orbit.

Then (in 1803) William Herschel discovered the orbital motion of binary stars – although the objective of his observation was still the stellar parallax. Herschel's discovery proved the validity of the law of universal gravitation in great distances from the Sun and Earth. This discovery went a long way to strengthen the belief in the universality of the laws of nature.

Not until the early 19th century, that is 250 years after the death of the originator of the heliocentric theory, did astronomers finally succeed in measuring the parallax of a few of the nearest stars. Success was achieved at a moment when no one had any doubt regarding the validity of the Copernican idea and of the improvements upon that idea made by Kepler, Galileo and Newton. The measurement of stellar parallaxes furnished astronomers with a method for measuring the distance of stars. That method still constitutes the basis for all linear measurements in our nearest and most distant environment.

A great deal has changed in astronomy since the days of Copernicus but much still remains unchanged. We still consider, as did the recluse in the Frombork tower, that the basic criterion of all theories is consistency of observation with conclusions derived from them.

The publication of *De revolutionibus* coincided with the death of Copernicus. The long and anxiety-filled life of Copernicus came to an end, but the drama for which he set the stage continues to this day.

Although the heliocentric idea has become universally accepted, although it is now a truism, a commonplace evoking little or no excitement, yet the background of that drama, the ceaseless struggle for a constantly new vision of the world and the effort to mold life to that vision, still constitute the setting for the dramatic clashes and fierce conflicts observed in our modern world. That is why it is worth our while to turn our minds occasionally to those distant times and to those forgotten figures. Having blown away the dust, we may observe in their faces the traces of the same hopes and passions that we know today, and read of events which, though they will never be repeated precisely yet will occur again and again.

FROM COPERNICUS TO EINSTEIN*

LEOPOLD INFELD †

> Summo erat ingenio non in rebus tantum mathematicis, verum in physicis aliisque omnibus...

The work of Copernicus represents not only the beginning of modern astronomy but also the beginning of the modern science of the universe and of nature. Awareness that nature is governed by laws emerged through study of the heaven and through observation of the Sun, the stars and the planets, for it is there that the laws are most striking.

We know from the history of science that the first step is always the most difficult. Copernicus made the first, the most difficult and the most significant step in the science of the universe. He transferred the frame of reference from the Earth to the Sun, proclaiming to the world that the planets revolve about the Sun and not about the Earth. Today we are able to range far and wide over the universe, we can project our thoughts not only from the Earth to the Sun, but also from the Sun to more distant stars, from the stars to the center of our Galaxy and from our Galaxy to the distant nebulae. In this journey across the universe, in which we can traverse millions of light years in thought, the first step, as I have said, was the most difficult. That first step was made by Copernicus.

In the 16th century, when the world was emerging from ignorance, when the Renaissance reigned in Italy, the prejudices of the ancient world and the benightedness of the Middle Ages still weighed upon scientific thinking. To assert that the Earth and the other planets move about the Sun required great scientific imagination, great scientific courage and, as it turned out later, great personal courage.

My purpose here is to describe the influence Copernicus exercised on the subsequent development of science and principally on the development of mechanics and the theory of gravitation. I should like to point out how the ideas of the following centuries, which had their beginning in the concepts of Copernicus, were in some ways far from them and in others reverted to them. Quite understandably, I can do no more than indicate a few of the major signposts along the way of this development, those signposts which I consider most significant. We encounter here only a few names – those of Kepler, Galileo, Newton and Einstein. Their names mark the major developments leading from Copernicus to the theory of gravitation as it is formulated today.

At the time of Copernicus, antiquity had bequeathed two principles to the world. The first stated that the Earth is stationary while all the planets move arround it, the second that the most perfect motion is uniform motion on a circle. It is interesting to

* This is a shortened version of an article published in Polish in *Mikołaj Kopernik. Szkice Monograficzne* (Nicolaus Copernicus. Monographic Studies), Warsaw 1965.

note that the revolutionary idea put forth by Copernicus liberated science from the first prejudice but failed to liberate it from the second.

Did Copernicus prove that it is the Earth and not the Sun that moves? Let us for a moment assume the position of a 19th century physicist and consider what in his view would constitute such proof. His reasoning would take the following line: We know that the Earth performs two motions. The first is the diurnal motion around its axis. We wish to demonstrate that we experience the sequence of day and night due to that motion and not due to the motion of the Sun. The Sun rises in the east and sets in the west. Consequently, the Earth moves from west to east. If we drop a stone straight down from a high tower, the stone takes part in the revolution of the Earth. Consequently the velocity of its west to east motion is greater than that of the Earth. On the strength of the principle of inertia, the rock arrests its velocity. Hence, falling to the ground it will be deflected eastward. The path it travels will not be a straight line but a line slightly deflected to the east. Let us now examine that statement. We know that it is based on the principle of inertia which was not known in the early 16th century although Leonardo da Vinci, a contemporary of Copernicus, had already anticipated it. Had Copernicus succeeded in performing such an experiment, he would have proved by it the principle of inetria, that is the fact that the Earth revolves about the Sun. Such proof would have had meaning only within the framework of dynamics which a 19th-century physicist would have understood thanks to the works of Galileo and Newton.

How can the annual motion of the Earth be discovered experimentally? The answer is simple. Imagine two stars, one near our Earth the other far from it. On the first of June (let us assume), we measure the angle between the two stars. We repeat the measurement on December first. The angle should change due to the Earth's motion in the ecliptic. This would provide proof that it is the Earth and not the Sun that moves. Copernicus searched for this phenomenon without success. However, he believed so strongly in the motion of the Earth that he finally arrived at the conclusion that the stars are too far away for this phenomenon to be discovered. Indeed, the phenomenon could not be observed with the primitive instruments which the humble canon had at his command.

Copernicus was left with only one argument. That is that the hypothesis about the motion of the Earth was simpler than the geocentric hypothesis. Simpler particularly in the description of the motion of the two planets Mercury and Venus.

We often find in science that two rival hypotheses vie with each other. A classical example of this are the two hypotheses on light: Huyghens' and Newton's. For Huyghens the propagation of light consisted of a wave motion in the ether; for Newton it consisted of the motion of weightless light corpuscles. Two theories which described the then known phenomena of light in the same way. At first Newton's theory won recognition thanks to the great scientific authority he commanded, although it explained known phenomena in a more complex manner than did Huyghens' theory. Later, however, the phenomenon of light interference was discovered and it could be explained only by Huyghens' theory. Consequently, the 19th-century physicist rejected the Newtonian theory and accepted Huyghens' theory. We shall not complicate this simple and edifying story by extending its history to the 20th century, but shall leave it at that.

The Copernican theory met with a similar fate. In his case, however, the decisive role was played by the authority of the Church which claimed that the geocentric theory was supported by religious doctrine. Yet here too, the steadily growing number of pheonomena which could be explained only by the Copernican and not by the geocentric theory, compelled 18th-century astronomers to accept that theory.

In the 16th century it was still possible to accept both theories. To us the geocentric theory seems absurd, even to people who know nothing about the stellar parallax and the deflection toward the east of bodies falling from a height. For we are brought up in the heliocentric system just as our ancestors were brought up in the geocentric system. The greater the admiration we should have for those who were able to defy old prejudices. *De revolutionibus* is preceded by an unauthorized preface from the pen of Osiander. Various authors have heaped invective upon the head of Osiander for this. That may be very well. But contrary to many opinions on that subject, the preface seems quite sensible. In the dispute that broke out, Osiander was accused of representing the Copernican theory as one of the possible hypotheses whose virtue lay in its greater simplicity. Copernicus, on the other hand, honestly believed that the Earth moves. But in the 16th century, the dispute was only about words. Every scientist who formulates a new hypothesis believes that he is describing the reality that surrounds us, although this fact may be discovered only subsequently with progress in science. We cannot reconcile the stellar parallax and the eastward deflection of a falling stone by the geocentric theory. But this belongs to the later history of the Copernican theory. Copernicus imagined that the planets revolved about the Sun in a uniform motion along the circumference of a circle. From the present point of view, it seems that it would have been much simpler to get rid of the prejudice that the planets move along the circumference of a circle than that the Earth forms the fixed center of the universe. In order to remain in agreement with observation, Copernicus had to disfigure his theory with motions in epicycles. About eighty years elapsed before science rejected the second prejudice, namely that planets move in a uniform motion on the circumference of a circle around the Sun. The man who laid this prejudice was Johannes Kepler.

1. Johannes Kepler

Born in 1571 in Würtemberg, he studied at the University of Tübingen from 1589, where he became acquainted with the Copernican theory. At first he planned to study Lutheran theology but, luckily for the development of astronomy, he took the post of provincial mathematician in Styria and devoted his entire life to science. Although this is not the place for a detailed account of his life, one cannot help but mention the fact that it was a tragic life passed in constant battle against poverty and even destitution, a life fraught with family trouble alleviated only by fruitful scientific work. His is a tragic but a generous and noble figure. His writing does not emanate the bitterness of a forgotten great astronomer.

Kepler's first work appeared in 1594 under the long title: *Prodromus Disertationum Cosmographicarum continens Mysterium Cosmographicum.* He is one of but a few astronomers who boldly stood up to defend the Copernican theory. He listed a series of reasons which demonstrate why the Copernican theory is more valid (that is more serviceable) than the Ptolemaic theory.

Throughout his scientific career his interest was focused on the following problem: we know that such and such are the distances of the planets from the Sun. Since Kepler believed in God, he was fascinated by the thought that God should have selected these and not other numbers. Guided by religious fervor, he erred in making statements which are absurd from the point of view of a contemporary physicist, only to discover the right law at the end of his life.

It is this problem which preoccupies him in *Mysterium Cosmologicum* (1596). He believed in the importance of his discovery which united in the most incongruous manner the distances of planets with regular solids. Nevertheless twenty-two years elapsed before Kepler discovered the true law that governed these distances. The second problem to which Kepler applied himself was the form of the planetary orbits. But measurements were highly inaccurate in his day. The most accurate measurements at the time were made by Tycho Brahe, a gruff, apodictic man who had in 1600 left his native Denmark and moved to Prague. Kepler also repaired to Prague where he worked at first as assistant to Tycho Brahe and later succeeded him as 'mathematician of the empire'.

In his subsequent works – *Progymnasala* and *De stella Martis* – he wrote on the shape of the orbits, basing himself on observations conducted by Tycho Brahe. Among many foolish statements, we find a genuine jewel – Kepler's second law which states that the straight line joining the Sun to the planet (the radius vector) sweeps over an equal area in equal intervals of time. This is not exactly how Kepler's law was worded by him; furthermore he committed two errors in his reasoning which cancelled each other out, but he arrived at this law after long study. He wrote:

> God in His bounty gave us a very diligent and accurate observer in the person of Tycho Brahe – it is therefore right on that basis to discover with a grateful heart the true motions of celestial bodies.

Kepler did not yet know his first law and continued to imagine that the only motion possible is motion on the circumferences of circles and on epicycles. Having found many inconsistencies with observation, he concluded that the orbit must be oval. But this theory was also unsatisfactory. He wrote:

> I woke as if from a dream and saw the light.

That is how Kepler's first law originated. It says: planets move in ellipses and the Sun is in one of the foci. It is interesting to note that the second law preceded the first in time of discovery. The law undermines the influence of the Middle Ages, the influence of Aristotle to which Copernicus yielded, convinced that the planets can move only on the circumference of a circle.

Finally the third jewel. In 1619, in his *Harmonices Mundi libri*, Kepler solved the problem which preoccupied his thoughts for over twenty years. In the terminology of his day it states: the cubes of the mean distances of the planets from the Sun, divided by the squares of the periods of revolution, are equal for all the planets.

The experiments and the computations Kepler had to carry out before he discovered that law!

Kepler died in abject poverty in 1630 yet it was he who finally liberated astronomy from medieval obfuscation.

Anyone who has taken an elementary course in physics in a secondary school or in the first year at the university knows Kepler's three laws. They invariably strike every student as dogmatic, detailed and dull. That this should be the case – and indeed it is – is because there must be something wrong with the way we teach and because we rob the student of that which is most important – his ability to wonder. Every scientific discovery is like a tree with its roots in the past and its crown in the future, or to use another simile, like a link in a chain. The role of a discovery cannot be understood unless the link is attached to the preceding one – the past – and to the next one – the future. Kepler presented the three laws in a fresher and much more exciting manner than does the dogmatic formula which we learn at school. With what great pride and imagination he wrote of his discovery, whose greatness and importance he fully appreciated:

> The aim for which I worked with Tycho Brahe the best part of my life will finally see the light of day. Nothing holds me back... Alea iacta est. The book is written and I do not care whether it will be read now or by posterity. It may wait a hundred years for an observer.

2. A Glance at Kepler's Dynamics

The work of Copernicus and of Kepler represents a single stage in the development of science. In that period both Copernicus and Kepler dealt with the description of the motion of planets around the Sun. The description of motion is indeed the first step in the history of its understanding. The second step consists in the formulation of the laws of motion on the basis of which we are able to deduce motion and its description. Thus, in the first period of scientific development we describe motion; in the second we deduce the character of that motion from the laws of physics. On the whole the description of motion does not enable us to predict the future of a system or to discover its past. Only laws enable us to predict the future of a system. As a rule mechanics, the study of motion, the basis of all the natural sciences, is divided into kinematics, a kind of introduction to mechanics, and dynamics, the study of the laws that govern motion. We might say that in the age of Copernicus and Kepler science was confined to kinematics, that is to say, to the description of motion. Not dynamic but only kinematic concepts appear in astronomy. That the kinematic approach, that is the description of motion, is alone adequate for predicting phenomena in astronomy can be explained by the fact that the motions we are describing are periodic and recurrent; that is why the observation and description of past occurrences enables us to predict the future.

In ancient times, in Aristotelian physics, it was assumed that velocity is related to force, that the greater the force acting on it the faster a body moves. A cart pulled by four horses moves twices as fast as a cart pulled by two horses.

Kepler failed to rid himself of that error. He sought a force that acted in the direction of velocity, that is in the same direction as the path of motion. That force, Kepler believed, is inversely proportional to the distance and has some connection with the fact that the Earth, like every planet, is a magnet. I shall quote what he wrote in the original because I feel that it would be difficult to otherwise convey the idea of his style.

Für mein Person, sage ich, dass die Sternkugeln diese Art haben, dess sie an eunem jeden Ort dess Himmels, da sie jedesmals angetroffen werden, stillstehen würden, wenn sie nicht getrieben werden solten. Sie werden aber getrieben per speciem immateriatam Solis, in gyrum rapidissime circumactam. Item werden sie getrieben von Ihrer selbst eygnen Magnetischen Krafft, durch welche sie einhalb der Sonnen zuschiffen, andertheils von der Sonnen hinweg ziehlen. Die Sonn aber allein hat in Ihr selbst ein virtutem animalem, durch welche sie informiert, liecht gemacht, und wie ein Kugel am Drähstock beständiglich umgetrieben wirdt, durch welchem Trieb sie auch Ihre speciem immateriatam ad extremitates usque mundi diffusam in gleicher Zeit herumb gehen macht, und also successive alle Planeten mit herumb zuecht. Mahrere scientia animalis wirdt zu den himmlischen bewegungen nicht erfordert.

Again (from the point of view of the 19th century physicist) this is pure nonsense. Mankind had to wait another fifty years before it was to shed the bonds of Aristotelian dynamics. Meanwhile, thanks to Copernicus and Kepler, kinematics was liberated....

3. The Kinematics of Copernicus and of Kepler and Newtonian Dynamics

The dynamic approach to the question of the motion of celestial bodies began with Galileo and Newton. What is the nature of the problem to which a partial solution is found in Galileo's work and a full solution in Newton's work? Let us imagine the planets Mercury, Mars and the Earth revolving about the Sun. What forces compel the planets to circle around the Sun. The formulation and the answer to that question was feasible only when the concept of acceleration and the fact that force is related to acceleration were explained – a fact which Galileo already understood when the motion varied uniformly, but which was not fully understood by anyone before Newton.

Two thousand years had to elapse before science, thanks to the experiments of Galileo and Newton, finally freed itself from the error that not velocity but acceleration is proportional to force. This is the beginning of mechanics, the beginning of physics. But let us return briefly to the Copernican problem. For the sake of clarity let us forget for the moment about Kepler's discovery and imagine, as Copernicus wished, that a planet, the Earth for example, moves in a uniform motion about the Sun. According to ancient concept, the force should act in a direction tangent to the circumference of the circle, that is in the direction of velocity. The same applied to motion along an ellipse – i.e., that the force should operate tangent to it or in the direction of the velocity. But we know that in uniform motion around a circle, force acts in the direction of the center of the circle. The word acceleration is not the happiest choice in this instance; the point is that in a uniform motion on the circumference of a circle velocity changes as a vector because the direction of the velocity changes even though its absolute remains the same.

When we have a uniform motion in a straight line then there is either no force or all the acting forces neutralize each other. There is no force because there is no acceleration and force, according to Newton's law, is proportional to acceleration. But when we have motion at a constant velocity along a curve and not in a straight line then we have acceleration, we have force, because the velocity changes; it changes its direction although it does not change its absolute value. Not until the 17th century, that is 140 years after the death of Copernicus, did scientists begin to understand that the direction of force must be the same as the direction of acceleration, that in the simplest case of the motion of the planet the force acts along the circumference of a circle; not in a direction tangent to the circumference of the circle, but in the direction of the radius, in the direction which joins the planet to the Sun. In the case of motion along an ellipse, acceleration, hence also force, is directed toward the Sun in the direction of the radius vector which joins the planet to the Sun. As a stone dropped vertically does not fall in a straight line but on a parabola, so in the case of a stone thrown horizontally the direction of velocity is not the direction in which the force acts. Similarly in the case of the planet, the direction of velocity is not the same as the direction in which the force acts. Indeed, we know that when we shoot a rocket with sufficient velocity, we can create an artificial planet, not despite the force of the Earth's attraction, but because of it and because of the great initial velocity in a direction opposite to the direction of the force of the Earth's attraction.

It required tremendous scientific imagination to understand that the motion of a stone falling freely to the Earth, that the motion of the Moon around the Earth and the motions of the planets around the Sun, that all these motions in the universe are governed by one single law – the law of universal gravitation. Any two bodies in the universe attract each other with a force proportional to the product of their masses and inversely proportional to the square of the distance between them. The story of the falling apple which gave Newton the idea of universal gravitation is of course pure fiction. Actually Newton had thought on the subject for fifteen years before he was able to prove his theory by taking the simple example of the Moon's motion around the Earth on the circumference of a circle. Very frequently the light of understanding dawns suddenly. But this comes about only when preceded by months and even years of constant thought.

Newtonian mechanics sheds light on the Copernican and Keplerian problem of the motion of planets around the Sun. Dynamics, the formulation of the physical laws that govern this motion, was born. Deduction replaced description of motion, marking an important step in the history of our knowledge.

In accordance with the principles of Newtonian dynamics, in accordance with the simple law of gravitation, we formulate equations of motion for planets revolving about the Sun. The equations of motion are second degree differential equations.

When we solve them, we arrive by pure mathematical means at the conlusion that the planets revolve about the Sun along an ellipse. Kepler's famous first law thus becomes a conclusion from Newtonian principles. But the great success of the theory lies not in this alone. It appeared that not only Kepler's first but also his second and third laws are consequences of Newton's law. Newton's law goes much further than Kepler's law because Kepler's laws result from Newton's. Suddenly reality becomes simple to us, even though we must first master the art of deduction, formulated by the theory of differential equations, before we can understand it. To every man of science reality seems understandable if he can deduce from simple principles new and frequently complex facts. It seems simple even though the chain of deduction is long. The chain of deduction increases commensurably with the development of science. Comprehension of this fact is essential to an understanding of the spirit of science. This is clearly evident in the case of Kepler and Newton and in their mutual relations. Kepler's laws seem artificial, incomprehensible and are difficult to remember. Kepler stated that planets move along the ellipses because he wished the observations to conform to some kind of curve. After a great many experiments he hit upon the fact that the curve is in the form of an ellipse and that the Sun is at one of its foci and not, as he believed earlier, at the center. Kepler's remaining two laws are even more abstruse and Kepler arrived at them by way of many tedious experiments. All the laws are the consequence of Newton's theory. Bur more than that! It appears that the laws are only the first approximation. The laws that result from Newtonian physics are much more profound and provide a better description of reality than those formulated by Kepler. This fact also calls for some explanation.

Newton formulated the laws of mechanics in three basic principles of mechanics. It is not so simple to gain a deeper undertsanding of these principles although every secondary school student knows them by heart, verbally at least. Furthermore, Newton formulated the law of gravitation.

One of the basic difficulties to be found in Newtonian mechanics is the concept of the frame of reference. We must discuss this subject at greater length here because, on the one hand, the concept is related to the Copernican theory and, on the other, it has a bearing on the progress of science. When we formulate Newton's laws the question that immediately arises is: to what frame of reference do these laws apply? Let us take a simple example: in a room with a smooth floor we push a ball lightly. We know that the smaller the force of friction the more uniform the motion of the ball. Newton's conclusion that the ball's motion would be uniform if no forces are acting upon it is in its essence an idealization of reality. But let us imagine that we repeat the experiment on a fast moving carousel. We then obtain a different result. The ball will try to move away as far as possible from the fixed point of inertia through which passes the axis about which the carousel rotates. We thus see that Newton's first principle, the principle of inertia, applies (more or less) to my room but that it does not apply to the carousel. But according to Copernicus, the Earth is a carousel rotating around its axis in the course of a day while at the same time it revolves about the Sun in the course of a year.

We ask ourselves: where is the frame of reference to which Newton's laws of dynamics apply? Certainly, the carousel is not this frame of reference, hence my room is equally not that frame of reference because, as Copernicus says, the Earth also revolves. Wherever a new difficulty arises new terms also appear. The new term in this case is the 'inertial' frame of reference. That is the name we give to a frame of reference to which the laws of dynamics applies. We might call the Copernican system the frame of reference connected with the Sun; that is we say that we are using the Copernican system when we describe motion in relation to the Sun, when we accept that the Sun is motionless and the the Earth revolves about the Sun. The frame of reference for the Earth we shall call the Ptolemaic system, that is a system in which we assume that the Earth is at rest while the Sun and other planets revolve about it. The greatness of Copernicus lies in the fact that, contrary to the opinion of the entire scientific and non-scientific world, he had the courage to proclaim and to prove that his system gives a better description of reality than does the Ptolemaic system, that the transition from the Ptolemaic to the Copernican system represents a major step forward in the history of science. Many years after his death, Copernicus had the entire official world of science, the Catholic and the Lutheran churches, ranged against him. It is not possible to fully appreciate the greatness of Copernicus without a knowledge of the scientific atmosphere that prevailed in his time.

Newtonian mechanics, the work of Galileo and Newton, sheds new light on the nature of the discovery of Copernicus. Using the terminology that was developed after Newton and thanks to Newton, we may say that one of the features of the greatness of Copernicus' discovery lies in the fact that he demonstrated that the Sun rather than the Earth is the inertial frame of reference, that the laws of mechanics relate to the Sun rather than to the Earth. Indeed, when we relate the motion of the planets to the Sun, when we apply the principle of mechanics to the frame of reference connected with the Sun, when we apply the law of universal gravitation to that frame of reference, then in conclusion we obtain Kepler's laws. There are, of course, many cases where we can consider the Earth as the inertial frame of reference, that is in those cases and for those phenomena for which the revolution of the Earth is of minimal importance as, for example, when we investigate the flow of current or of electrons. But in gravitational phenomena, where the description of the motion of the Earth is the point at issue, there the frame of reference connected with the Sun rather than the frame of reference connected with the Earth appears as the inertial frame of reference; there the laws of dynamics refer to the Copernican system rather than to the Ptolemaic system. I have used the word 'rather' several times. Does that mean that the Copernican system is not strictly inertial? The answer to that question is yes. The Copernican system is not strictly inertial. The Copernican system is not the one to which the laws of mechanics strictly refer.

It would appear, therefore, that Newton's theory repudiates the Copernican theory. This kind of conclusion would be premature and superficial. Every theory is limited in its application. As progress is made in science, every scientific theory ultimately proves to be only an approximation; it calls for changes that are at times evolutionary and at times revolutionary. The greatness of the Copernican theory lies in the fact that there is no return from it to the Ptolemaic theory and, as demonstrated by later historical developments, in the fact that the Copernican ideas formed a fundamental change which staked out the direction for the further development of physics.

A better frame of reference that is more inertial than the one connected with the Sun is the frame of reference connected with the center of mass of our planetary system. Let us imagine that we have in our universe only two bodies of equal mass. The center of these bodies will then lie at the center of the line joining them. When, however, the first mass is a thousand times as great as the second then the center of mass lies a thousand times closer to the greater mass. Similarly, there is a center of mass for our solar system. That center of mass lies in the Sun (although not in the center of the Sun) because the mass of the Sun is very great in comparison to the total mass of the planets. It follows from Newton's mechanics that the center of mass rather than the center of the Sun constitutes the inertial frame of reference or in other words: the system connected with the center of mass rather than the system connected with the Sun is the inertial frame of reference. We have already said that Kepler's laws follow from Newton's laws as the first approximation. In reality, Kepler's laws are not true in the strict sense. They are true for our solar system with a great degree of approximation because the mass of the Sun is very great in comparison to the mass of the planets.

Let us look at the problem from a slightly different point of view for a moment. When the mass of the Sun is very great in comparison to the mass of a planet (and if we had only the Sun and the planets in our system) then we may understand the motion of a planet as a one-body problem. We say that the Sun is at rest while the planet moves around the Sun along an ellipse. We may do so because the center of mass lies practically in the center of the Sun. But let us imagine that the mass of the Sun is only several times as great as that of the planet. According to Newton's theory, we must relate the motion to the frame of reference connected with the center of mass. Then both the Sun and the planet, would move along ellipses and the center of mass would be at a focus. We thus see how Kepler's first law, like his next two laws, become changed. The frame of reference connected with the center of the Sun must be replaced by a frame of reference connected with the center of mass. For the sake of brevity we shall call the frame of reference connected with the center of mass the Copernican system (like the frame of reference connected with the center of the Sun when we assume that the mass of the Sun is very much greater than the mass of the planet).

The Copernican system is certainly more inertial than the Ptolemaic system. But we see in the development of physics up to Newton's time – and even more clearly later – that the Copernican system is not ideally inertial. Indeed, what we have in mind when we say that the entire planetary system revolves, that our entire Galaxy revolves, is precisely that the Copernican system is not strictly inertial. No revolving system can be strictly inertial.

Even this brief discussion points to the essential weakness of classical mechanics. The question arises: where is the inertial system of reference? What in nature, what in our objective reality, corresponds to the inertial frame of reference? Classical mechanics is not capable of providing an answer to this question. It tells us that system *A* rather than system *B*, that is the Copernican rather than the Ptolemaic system, is more inertial. But it does not tell us that this system is the inertial frame of reference.

Let us imagine that we have a system which we may consider as the inertial frame of reference. Then, according to classical mechanics, every other system moving in relation to that given system in a uniform motion is also inertial motion. That means that any other system which moves in a uniform motion in relation to the Copernican system is as good as the Copernican system. In classical mechanics, therefore, we are always dealing not with one but with a group of inertial frames of reference. We have been perhaps too technical and pretentious in stating the extremely simple fact that the same laws are in force in a train as on the ground, if the train moves in an ideally uniform motion in a straight line. The same law of mechanics is in force in all systems if they move in a uniform motion in relation to each other. What happens when the systems move in an irregular motion in relation to each other? Let us go back once again to the carousel where the principle of inertia is not in force. What laws are in force there? We may reply briefly: the laws formulated by Newton are in force only in an inertial system whereas other laws are in force on a carousel, for instance, laws that we are also able to formulate if we know how the given system moves in relation to the inertial frame of reference. New forces, known as D'Alembert's forces, come into play. Sometimes we speak of apparent D'Alembert forces although I doubt whether the term 'apparent' is the correct one.

Let us now briefly recapitulate what we have said above. The great feat of Copernicus marks the beginning of contemporary science. The remarkable discoveries of Kepler change the Copernican motion of planets in circles to motions on ellipses. Physics, and specifically the mechanics formulated by Galileo and Newton, enable us not only to describe motions but also to predict them. Kepler's laws derive from Newton's law of gravitation. The gravitational force of attraction acts on the rate of acceleration and not on velocity. It appears from classical mechanics that the Copernican system is, in regard to the motion of planets, an inertial frame of reference although later, with the development of scientific knowledge, it was found that this system is not inertial. By the term 'Copernican system' we should understand the system connected with the center of mass of the planetary system rather than with the center of the Sun, although the difference is not important because the mass of the planets is very much smaller than the mass of the Sun.

4. The Copernican Theory and Einstein's Theory

So far, we have been looking at the work of Copernicus, Kepler and Newton through the eyes of a 19th-century physicist. Why a 19th century-and not a 20th-century physicist? Because the astronomer (or physicist) of the 20th century knows the theory of relativity and his views are therefore basically different from those which characterize the 19th-century physicist. Indeed, in the current century even the Church has become reconciled to the Copernican theory, having taken it off the index of banned books in 1822.

We have already discussed the influence of the Copernican truth on two fields: kinematics and dynamics. We now come to the third and final level, to the description of the relation between Copernican theory and the theory of relativity.

A distinction is usually made between the special, or restricted, and the general theory of relativity. It is far more difficult to understand the general than the special theory of relativity. But as far as the Copernican problem is concerned, as regards the theory of gravitation in its connection with the Copernican theory, we are primarily interested in the general theory of relativity, which was formulated by Einstein, gradually and not without initial errors, from 1911–1916. The theory represents the first attempt since Newton's time to revise the question of gravitation. The root cause of this theory, as of every theory of physics, is the desire for a better understanding of the reality around us; in this instance it is the desire for a better understanding of the phenomenon of gravitation, the phenomenon of the motion of celestial bodies.

We have seen that Newton's theory introduces a very important concept, that is the concept of the inertial frame of reference. The Copernican system is more inertial than the system connected with the Earth; there is no truly inertial system in existence and the term should be abolished from science altogether. A theory which does not employ the term inertial frame of reference would be simpler than one based on the concept of that frame of reference, to which nothing in reality corresponds. But we must create a new theory if we are to rid ourselves of that concept. The theory of relativity is that kind of theory. Although it is simpler in its logic, the chain of reasoning that leads from observation to the description of our world is considerably longer. As usual, assuming less we must deduce more. The same was true earlier in the 17th century when we deduced Kepler's laws from Newton's theory as approximate laws.

Clearly, however, the simplicity of a proposition is not the only factor determining the importance of a theory. The evolution of science constantly demonstrates that every theory has a limited scope of applicability. In the eyes of a research scientist, it is increasingly more difficult to describe reality. Progress in science reveals the limits of applicability of earlier theories. They remain true, but only within the bounds described by the development of physics. We look for new theories which will take account of new experimental data, which will give a better description of reality than earlier theories. And again, we discover that the new theories are also limited in their applicability and so we are again forced to reach higher to new theories, to a new description of the world around us. And so ad infinitum through new discovery, errors, dissapointments and again new discoveries, we seek to discover absolute truth, the ultimate description of our material world, knowing full well that this is a mirage that can be attained only within specific limits, but attainable in fact with ever closer approximation. The goal of the human race is thus to gain knowledge of our material world, knowing that it is a mirage attainable only within certain limits but attainable in fact with ever closer approximation. Thus the goal of the human race is to gain knowledge of our material world, to discover the laws that govern it. Another equally important goal is to make use of these laws for the benefit and the happiness of mankind.

But let us return to the theory of relativity. Like every other theory, it must be a signpost in an area where the earlier Newtonian theory fails. We shall see later whether and to what degree the theory of relativity performs this function.

Our aim here is not to explain the theory of relativity but to explain its relation to the Newtonian and the Copernican theories. That is why we shall concentrate on the discussion of only a few selected aspects of the theory.

We must leave for a moment the consideration of the theory of relativity and interject parenthetically a reasoning which has a very important bearing on our subject. It concerns the structure of mathematical physics. The reason for our remarks is the fact that a philosopher often finds it difficult to understand certain questions that are obvious to a physicist. Cooperation between physics and philosophy calls for a calm and factual explanation of these problems. It seems to me that what I shall say here will find approval in the eyes of every theoretical physicist. It is of utmost importance to me to set right a certain misunderstanding that frequently crops up in the literature.

We must make a distinction between the mathematical structure of physical law and its physical content. Misunderstanding is often caused by the failure to make this distinction between the two important but entirely different concepts. We can explain the difference by the following example. We take a simple mathematical equation which describes the motion of a pendulum with a small deflection. An equation of the same mathematical structure describes the vibrations of a tuning fork, the vibrations of electric current and the vibrations of electrons. It would be absurd to say that the pendulum is the same as a tuning fork. The mathematical structure tells us nothing of the fragment of reality it describes until we relate the mathematical symbols in the equation to reality. Symbols that appear in mathematical formulas represent certain magnitudes in our real world. Until we establish what they represent, our theory is an abstract theory; it has a specific mathematical structure but it has no physical content. This does not mean that mathematical structure does not interest us; on the contrary, it interests us very much because once we know well the mathematical structure of a theory (i.e. when we know the solution of some differential equation), once we know what that fragment of reality is, we are then able to quickly draw conclusions from that theory by deduction and are again able to translate these relations into a language describing the given fragment of reality. Mathematical equations related to reality combine to give us physical theory.

This kind of theory gives an accurate description of reality if the conclusions, derived on the mathematical plane and then interpreted in terms of physics, agree with experience. Theory then represents a good description of our reality until further research indicates that the theory leads to contradictions with reality and that it must be changed. Hence, every physical theory must be characterized not only by mathematical structure but also by relevance of the given mathematical structure to the fragment of the material world to which it refers. To give a specific example: the principles of Newton's mechanics may be formulated in the form of equations related to a given frame of reference. But mathematical relations become physical theory when we know that m stands for mass and f for force. Newton's equations relate to a given frame of reference. In mathematical structure, a frame of reference is represented by three axes perpendicular to one another. What in our physical world corresponds to these perpendicular axes? The answer is: the inertial frame of reference ,i.e. that material frame of reference which is connected with a material body. But where is that material body and where is the inertial frame of reference represented in our mathematical structure by the set of axes? Once again we return to the answer given earlier: as regards the question of the motion of planets, that frame of reference is the Copernican system, but as regards the question of the motion of the entire planetary system, we must look for the inertial frame of reference outside the Sun. We see, therefore, that it is possible to establish the relation between an abstract concept of the mathematical frame of reference and the physical inertial frame of reference only to some degree of approximation.

Does the same difficulty arise in the theory of relativity which describes the problem of gravitation? No, because the mathematical structure of the theory of relativity does not contain the concept of a mathematical frame of reference. The concept of a mathematical frame of reference is a matter of indifference to mathematical equations of the theory of relativity (I draw attention to the word 'mathematical'). There is absolutely nothing in the mathematical structure of the theory of relativity that corresponds to the difference between the Ptolemaic and the Copernican system, to that vast difference between the two systems whose understanding we owe to Copernicus and whose understanding marks the beginning of our science of the world. This is one of the differences between Newton's and Einstein's theory; in Newton's theory we had the concept of a system which, it is true, corresponded to an inertial frame of reference that does not exist in nature: here – in the theory of relativity or rather in its mathematical structure – the concept of the frame of reference does not appear. But, as we have said before, theory is comprised not only of the mathematical structure but also of its connection with reality. The criticism that the mathematical structure of the theory of relativity has nothing in it that corresponds to the difference between the Ptolemaic and the Copernican system is not valid because physical theory must be judged only when the mathematical structure of that theory is connected with reality, for only then is it a physical theory. In order to relate the mathematical structure of the theory of relativity to reality, I must first choose a given system in the physical world. For the description of the motion of planets the choice should fall on the Copernican and not on the Ptolemaic system.

In order to understand this, let us consider for a moment what questions we address to the theory of relativity. I shall take an example which is perhaps the most important one. According to Newton's theory, the planet Mercury moves in an ellipse around the Sun and the Sun is at the focus of that ellipse. Do we derive the same or a slightly different conclusion from the theory of relativity? First of all, we know that any deviation from Newton's theory must be very slight for otherwise Newton's theory would have been discarded a long time ago. If such deviations do exist, they will be especially clear for the planet Mercury because it is nearest the Sun. But let us turn our attention to our question again. We are looking for a difference between the prediction of Newton's mechanics and that of the theory of relativity. However, we shall be able to evaluate that difference only if we describe the motion within the same frame of reference. We must describe the same actual motion within the same frame of reference, once in accordance with Newton's theory and once in accordance with the theory of relativity. We must see whether there is any difference and establish whether observation proves or disproves the conclusion derived from the general theory of relativity. The theory of relativity predicts the Newtonian motion, that is the motion of the planet Mercury along an ellipse, but it also predicts that the ellipse revolves very slowly. This is what we call the perihelian motion, where perihelion means that point on the curcumference of the ellipse at which the planet is nearest the Sun. Since the planet revolves, that perihelian point revolves very slowly on the circumference of a circle. In the course of a hundred years the line joining the perihelian point with the Sun will revolve 42 seconds! This conclusion was remarkably confirmed by observation. The perihelian motion proceeds in the same direction on an ellipse as the Newtonian motion. There is very good agreement between the prediction of the theory of relativity and observation. But, as I have said before, in order to be able to compare the prediction with observation, we must relate motion to the Copernican system.

It is not true, therefore, that the theory of relativity can make equally good use of the Ptolemaic system because, as in the case of the motion of Mercury around the Sun along an ellipse, the question of the Earth does not appear at all. The calculation, which is fairly complicated, refers to the following problem: how does a very small body (Mercury) move in the field of a large and heavy body that is at rest (the Sun)? This is called the problem of the motion of one body in the field of a heavy body; a typical Copernican problem where there is no mention made of the motion of the center of mass because it is assumed that the Sun is infinitely heavier than the planets. The description is entirely objective because the fact that the observer is on the Earth finds expression neither in the formulas nor in the calculations. The problem of one body is fairly simple in the theory of relativity but it, too, presents difficulties in comparison with Newtonian calculation.

In Newton's theory the two-body problem, double stars for example, is not any more difficult than the problem of one body of small mass moving in the field of a body of large mass (Sun, planet). To all practical purposes, the calculation remains the same, except that the motion should be related to the frame of reference connected with the center of mass of those two bodies.

And how does this matter look in the general theory of relativity? Using the example of Mercury and the Sun we have already explained the problem where one light body is in the field of a heavy body. But in the case of two bodies whose mass is comparable, the calculation is vastly more difficult. Indeed, the problem has a long history; the solution to it appeared for the first time in 1938. The problem still constitutes the subject of many scientific papers. Here again we ask the following question: does the motion of two bodies predicted by the theory of relativity differ from the motion predicted by Newton's theory? And again the concrete answer is: when we solve the question according to the theory of relativity we accept the system connected with the center of mass of the two bodies, hence the same system that we accepted in the Copernican theory. And here again we find a deviation from the Newtonian motion, specifically the very slow perihelian motion.

Finally, we might mention one other example, that is the famous deflection of the rays of the Sun. A ray from a star bends when it passes through the gravitational field of the Sun. And again the deflection is related to the Sun and not to the Ptolemaic system. The Ptolemaic system plays no role in a theoretical analysis of these experimental facts. The description of these facts is entirely objective, independent of the observer. Obviously, however, these facts must be established by man on Earth. Man must use certain instruments: telescopes, theodolites and others. He must know certain additional laws in order to be able to evaluate the phenomenon. He must know the laws that govern the functions of his instruments and the laws that govern the properties of sight. Let us take a specific example in order to explain what I have in mind. Let us assume that the light of a star passes near the Sun. According to the theory of relativity, the ray of light is deflected. Now let us assume that the ray falls into our eye. We must know the physiological law which states that our eye sees the star on the tangential extension of the ray of the star that reaches our eye. The camera functions in the same manner. That means that a group of stars near the Sun photographed during an eclipse of the Sun gives us a slightly different picture than the same group of stars photographed on a clear night. That is because in the first case the rays from the stars are bent by the Sun, while in the second case they are not. This is, as we know, the famous confirmation of the theory of relativity which served to spread its renown all over the world.

We have inferred an objective fact from the measurement described, but the theory of measurement is a necessary adjunct to the theory of relativity, an adjunct which allows us to establish whether or not the description of reality is valid. The theory of measurement is based in every case on the following principle: that observations are made far from the Sun. We may assume that where observations are made the gravitational field is weak and that its geometry is Euclidean, that is that it is governed not by the general by by the special theory of relativity. The principle is evident in another phenomenon which can be proved experimentally: the spectrum of an atom emitted by the Sun is displaced slightly toward the red in comparison to the spectrum of the same kind of atom on Earth. That is so because the atom in the Sun is in a very strong gravitational field while the atom on Earth is in such a weak gravitational field that one need not take account of its influence on the spectrum. The Earth is here only a point distant from the Sun. Or: when the atom emitting rays is on the first floor up and the observer of this radiation is on the ground floor, then he will ascertain a shift to the violet when the source is found in a weaker gravitational field than the observer. The shift was discovered recently thanks to the Mössbauer effect and its quantitative agreement with the general theory of relativity was established.

Therefore, when speaking of the mathematical structure of the theory of relativity, then indeed the invariance there expresses that the concept of a frame of reference is not necessary and that there is no difference (I repeat from the mathematical point of view of the theory) between the Ptolemaic and the Copernican system. We have an entirely different problem as regards physical content.

A mathematical description of a concrete fragment of reality, such as the motion of a planet, the motion of two bodies, the curving of light rays, is entirely objective and refers either to the system connected with the Sun or to the system connected with center of mass, or to the Copernican system. As an instrument in our discovery of nature, the theory of relativity makes use of the Copernican system to exactly the same degree as of Newton's theory. It is superior to Newton's theory by the fact that the concept of the inertial frame of reference is not necessary. It is superior to Newtonian mechanics by the fact that its conclusions are in greater agreement with observation than the conclusions of Newton's theory. In the theory of measurement, which we must add to every theory, we assume that measurements are made by the observer far from the Sun and that for him the gravitational field is very weak. The Earth and the observer intervene only through the theory of measurement: we must have it at all times so as to be able to deduce from the results of the measurements the properties and laws that govern the objective world. It seems to me that there is no doubt whatever that the theory of relativity represents an enormous advance in our knowledge, that its physical content is in agreement with experience. It is a relatively difficult theory, requiring many years of thought and a knowledge of mathematical methods. But the old saying that only twelve people understand it is completely obsolete. There were 150 persons attending the Gravitation Conference in Poland in 1962, and they not only understand the theory of relativity but also work creatively in it. I do not know how many people know and understand the mathematical structure of the general theory of relativity, but I believe that there might be more than a hundred times twelve.

None of them believe that this theory detracts in any way from the contribution made by Copernicus. It expresses the Copernican idea differently, in more modern terms. It says that for the description of phenomena we should take a system in which the gravitational field disappears into infinity.

Lagrange said of Newton that Newton was not only the greatest but also the luckiest of scientists because the science of the universe can be created only once, and Newton created it. We know today that evolutionary and revolutionary changes occur in science at such a rapid rate that they are synonymous with the fact that science can be changed, transformed and created many times. Since 1543, that is since the beginnings of the science of the universe initiated by the great work of Copernicus, 430 years have elapsed. Humanity has unlimited time before it. But in those 430 years great changes occurred in our physical outlook on life. For a modern physicist, the universe appears entirely different. He sees completely different laws governing the universe than those that were recognized by physicists in Newton's lifetime. In the 19th century, the physicist could still believe that the science of the universe can be created only once, that the mechanistic picture of the world is given once for all times, that the future generations will simply fill in the details and that basically it will not change in character. We know today that this is not so. That is why Legrange's statement about Newton is not correct. One is more nearly correct when one says that Copernicus was the greatest and the luckiest of astronomers because the science of the universe can be begun only once. And this good fortune fell to Copernicus.

THE GREAT RENAISSANCE SCHOLAR

WALDEMAR VOISÉ

Faculty of History of Science and Technology of the Polish Academy of Sciences

> ... ad quae ingenii ornamenta praeclara sibi paraverat subsidia ex linquarum Latinae et Graecae perfecta cognitatione. In medicina velut alter Aesculapus celebratus ac si animo prorsus philosophico ostentationem aput vulgum nunquam affectaret.

1. The Crucial Year of 1543

Although the specification of certain dates as crucial to the history of one or another branch of science, or to history in general, might raise objections in some quarters, it is nevertheless difficult to resist the temptation to associate momentous events in the history of culture with specific dates, solely, however, for the purpose of orientation, that is in order to get a better grasp of the continous current of history.

The year of 1543 seems to be such a date. It is linked with many important events in the history of culture and science. That year witnessed a number of important occurrences in Poland and in Europe as a whole. Notable among these was the appearance of two books, namely, *De humani corporis fabrica libri septem* by Andreas Vesalius and *De revolutionibus* by Nicolaus Copernicus. It had been remarked quite a while ago that, as the work of Copernicus marked a breakthrough in our concept of the macrocosm – of the Great World – embracing the whole of our galaxy, so the work of Vesalius represented a revaluation of the microcosm, that is the study of the Little World, embracing the human organism.

That Vesalius took the step, which proved of critical importance to the development of medicine, is a long held opinion that might bear repeating here. He introduced a new concept of anatomy to replace Galen's system, which constituted the basis of medical studies until then. This statement alone does not convey the atmosphere in which the treatise on the structure of the human body was written and, it might be added, 'the case of Vesalius' was a fairly complex one.

Although Vesalius performed his dissections according to the program established by Galen and although he retained Galen's order in the treatise he wrote, yet his frame of reference differed very significantly. Galen dissected, not human cadavers, but dead monkeys and dogs and on the basis of analogy arrived at what were frequently false conclusions regarding the human anatomy. Since Galen assumed that the human organism represents a perfect form of the animal organism, he made no distinction between the animal and the human world. Vesalius thought differently. To him the human being was a creature which occupied a highly privileged position in nature .He was therefore deeply convinced that the study of human anatomy should be based exclusively on the study of the human species. He believed that comparison of animal and human anatomy is pointless, that it only serves to obscure the issue. The attitude, to which Vesalius owed his remarkable achievement, was based on a principle that was as traditional as it was false. Had it been embraced by those who followed him, medicine would have strayed from its true goal. While Galen subscribed to pagan animism, Vesalius professed the Christian dogma which declared that the world (together with the animals, plants and so on) was created for the benefit and the welfare of man, 'the king of all creation'. The question of man's position in the natural world ignited a controversy which grew in violence throughout the 16th century and which, when it subsided in the centuries that followed, assumed the character of a philosophical debate. The question asked was, is there only a quantitative or is there rather a qualitative difference between man and animal? In other words, do human beings and animals belong to two distinct worlds or does man differ from the animal solely by virtue of his more complex physical structure and higher level of intelligence? Calling man 'the little world' (microcosm), Vesalius ascribed to him a central position in the natural world and, upholding this conviction, rejected the idea of any analogy between the structure of the human and the animal anatomy. Whenever he did become interested in the anatomy of a dog or a monkey, his purpose was to point out the fundamental differences which separated the human and the animal worlds. Thus, readers of his treatise on the structure of the human body had before their eyes a 'homocentric' picture of the Earth they inhabited supplemented by a 'geocentric' picture of the Universe. Vesalius wrote, among other things, that the Earth remains motionless and that it is in the center of the Universe.

These particular views were not unusual in the Renaissance. On the contrary, they were typical of the times and were professed by the most critical and perceptive minds of the age. To deracinate the deeply rooted conviction regarding the structure of the Universe and man's place in it represented no mean feat. Copernicus was well aware of this fact and so for many long years he hesitated to disclose the bitter truth contained in the Books *On the Revolutions*. In this light, the fact that the book appeared in print in the year of the author's death assumes a singular poignancy.

2. The Background of the Copernican Discovery

The course which Copernicus elected to follow in order to arrive at the truth, identifies him with the Renaissance of which he was one of the most typical products. Although the concept of truth as a 'union of thought and object' was not formulated by Francis Bacon until the close of that era, yet many thinkers of the Renaissance adopted the principle that the two elements forming the foundations of human knowledge are the intellect and the object, or to put it differently, the discoverer and the object discovered.

Despite the various vacillations and despite the insinuations that were made at the time, the core of the problem was clearly defined. Sense perception and the intellect were conceived as factors whose interaction is indispensable to the advancement of knowledge. Experience and the mind, treated jointly as an instrument of investigation

– the time had not yet arrived when new scientific instruments (telescopes, microscopes and others) had been invented – nearly always signified a critical approach toward the dogmatic attitude. That is why the study of ideas advanced by others on the subject of reality slowly receded into the shadow, while the investigation of reality itself was set up as the goal. But, contrary to views which persist to this day by what seems like the sheer force of inertia, Copernicus does not owe his great discovery to thousands of observations of the heaven. He followed a different line of reasoning. His point of departure was the contradictions he discovered in the traditional Ptolemaic system and, as he writes himself, the inspiration to look for a new theory, which would offer a better explanation of daily phenomena, was provided by the works of ancient Pythagoreans who had in their time postulated the possible motion of the Earth's sphere.

Copernicus did not always have to go so far back in time to find material from which he could evolve his ideas because a number of remarkable thinkers and scientists closer to his time had cast doubt on the principles of Ptolemaic geocentricism. There was primarily the great French philosopher Nicole Oresme who advanced the hypothesis of the diurnal motion of the Earth in his treatise *On the Heaven and the Universe*, written in 1377. A hundred years later the German thinker Nicolaus Cusanus formulated the original concept that our motionless Earth does not remain at a permanent (fixed) distance from the Creator. Oresme and Cusanus played a major role as intermediaries between medieval natural philosophy and modern developments in that area: the new thinking also contained a fair dose of criticism of the ideas of Aristotle and Ptolemy. Greatest reservations were aroused by the ancient principles of physics which stated that the laws of motion in force on Earth did not apply elsewhere. Jean Buridan, rector of the University of Paris and a contemporary of Oresme, was chief among those who challenged the assumptions of Aristotelian physics. He propounded the principle that the same laws of the motion of bodies applied to the Earth and to interplanetary space.

Copernicus proceeded similarly in his time. With all the respect he had for the ancients, he did not hide the disillusionment he felt after having carefully perused their works. We know how distressed he was to discover that some of the ancient astronomers had deliberately distorted observed data in order to save at least the apparent validity of their theories. He had already heard in Cracow from the lips of his teacher, Wojciech of Brudzewo, that the conclusions of many of the successors of Ptolemy (including Peurbach, author of the astronomy textbook then in use) are not free of error and that one must not reject a priori the theory which states that not the Earth but the Sun is in the center of the Universe. Later in Bologna, Copernicus attended the lectures of Novara who proposed the adoption of a new system for measuring the distances of celestial bodies. Finally, it seems very likely that during his sojourn in Ferrara Copernicus might have personally met the humanist Celio Calcagnini, author of the treatise *On the Immobility of the Heaven and the Mobility of the Earth*. The preserved notes written in his hand on the margins of books which had once belonged to Copernicus would seem to point to the influence the early concepts of the astronomer (Copernicus) may have exercised on those ideas of the Italian humanist which came very close to heliocentrism.

It should be made clear, however, that all the concepts listed above were no more than elaborate assumptions on the subject of the structure of the Universe. At best they could provide Copernicus only a very general incentive to his own reflections on the subject. No scientific premises had been yet produced that would confirm the motion of the Earth, while the relation of the Earth to the Sun still remained within the sphere of speculation unsupported by objective and scientific argument. And this could be provided by someone who was, not only bold enough to accept a hypothesis that differed from the traditional one, but who was also capable of proving its validity in a manner that, based on the most precise observations possible, would meet the rigid demands of correct reasoning.

However, this was not as easily done as it is said. To present a revolutionary theory to the world and to convince the world of its validity presupposed not only a large dose of scientific imagination but also great erudition. In spite of its ancient lineage, the theory itself was so novel that it may be placed among the most crucial theories in the development of the human intellect. When at the close of the 18th century the great German philosopher Emanuel Kant wrote of the origins of his system, he noted that in formulating it he proceeded in the same manner as Copernicus "who, when the description of the motion of the celestial bodies, based on the principle that the whole array of stars revolve about the Sun, did not succeed, tried whether he would succeed better by stopping the stars and by having the observer revolve instead." These words illustrate the importance of the original concept which became the pivotal point of the Copernican system.

At first the heliocentric theory was no more than what is now usually called a working hypothesis. Copernicus devoted all the subsequent observations of his busy life to one purpose: to supply proof for what seemed like a paradoxical thesis which contradicted ordinary good sense. Although every new observation he made confirmed his earlier conviction, he nevertheless continued to keep a sharp lookout for any among the investigated cases which would contradict his hypothesis. This fact would seem to imply that he recognized sense perception, whose purpose is to confirm or to deny the validity of reasoning, as the ultimate and most important test of truth. Although he chose induction, that is reasoning from the particular to the general, as the method by which he sought to explain the laws that govern the motion of celestial bodies, yet he chose a different method of reasoning, to prove the validity of his thesis.

Copernicus wrote his great work, *On the Revolutions*, when the first elation of the discovery had worn off and when he had already dismantled the scaffolding raised around the structure he had erected by the power of his genius. We would have never learned what lies behind the simple, clear-cut and lucidly logical force of the figures and numbers had it not been for Rheticus, who was less classically objective than his great 'master preceptor' ('dominus preaceptor' is the term Rheticus used in reference to Copernicus). It is thanks to Rheticus that the poetic quality of the great Pythagorean's musical imagination had survived its originator. *To the famous Johannes Schöner, on the book of revolutions by the very learned and remarkable mathematician, the Honorable Doctor Nicolaus Copernicus of Toruń, Canon of Warmia, by a youth enamoured of mathematics, the First Account:* under this title there appeared in Gdańsk in 1540 a work by Rheticus, master of liberal arts, at one time professor of mathematics at Wittenberg. Among the arguments supporting the motion of the Earth listed in *The First Account (Narratio Prima)* is also one (the sixth) which Rheticus begins with reproof levelled against ancient astronomers. It says that they

fashioned their theories and devices with too little regard for the rule which reminds us that the order an motion of the heavenly spheres agree in an absolute system. We fully grant these distinguished men their due honor, as we should. Nevertheless, we should have wished them, in establishing the harmony of the motions, to imitate the musicians who, when one string has either tightened or loosened, with great care and skill regulate and adjust the tones of all the other strings, until all together produce the desired harmony, and no dissonance is heard in any.

The adherent of the Pythagoreans could not leave out this argument for it constituted one of the most significant links in the reasoning of philosophers who, as very few before and very few after them, were adept at combining the rigid demands of logical reasoning with a rich poetic imagination. And so we can picture Copernicus as one night in his observatory he ascertains that the motion of planets is governed by a different order than the one described by Ptolemy centuries earlier. The ancient Pythagorean music of the spheres, the symphony of the Universe, must have then echoed in his imagination. Having heard it when we are born and throughout our life and hearing it still even as we die, our human ears become deaf to this constant companion of our life in the same way as we do not feel the scent of the most aromatic perfumes when we smell them all the time.

Perhaps the first to draw attention to the 'musical' argument of Copernicus was Jeremi Wasiutyński. He wrote what follows in a book, very much in vogue before the Second World War, which continues to be the subject of a heated controversy:

Copernicus did indeed use the tuning method to establish the dimensions of planetary orbits. Neither Ptolemy nor his successors were able to do this for they did not possess the tuning fork, the fundamental tone, that is the orbit of the Earth which was discovered by Copernicus. Observing the dimensions of the arcs of retrogression described by the planets, caused by the revolution of the Earth, Copernicus, as the tuner does, listened attentively to the harmony produced by the planetary string with the string of the Earth. Mathematical calculation later gave him the relations of the orbits of Mars or Venus, Jupiter or Mercury to the Earth's orbit in the same way as the relation of the frequency of vibrations or the length of strings in a third, fifth or an octave are known.... The joy Copernicus experienced when he discovered the key to the heavenly lyre is reflected in the name he gave to the Earth's revolution when he called it the Orbis Magnus (the Great Orbit).

It was this 'musical' postulate which Copernicus, as a Pythagorean, took for his principal proposition. Pythagoras was the first to call the Universe an order (the Greek *kosmos*) and his number seven, manifest in the seven vowels, the seven musical strings, the seven stars of the Pleiades, was considered sacred. A hundred years after Copernicus, another great astronomer, Johannes Kepler (who perfected the Copernican system), also yielded to the temptation to seek a correspondence between the dimensions of planetary orbits and the musical intervals and considered it his signal triumph when musical intervals did indeed apply to the orbital system. Plato, who lived two thousand years before him, would have been jubilant at the fact that the distances of the Sun, Moon and the Planets have the same relation to each other as the double intervals, which he understood as the progression of whole numbers that are the powers of two and three. There must be something extremely fascinating about the Pythagorean concept for even in the 20th century James Jeans, when he wished to make the new discoveries in physics understandable, wrote that the laws to which the nature of the universe is subordinated recall not so much the principles that govern the motion of a machine as they recall the rules followed by musicians composing a fugue.

In the days of Copernicus, the Pythagorean concept of music had lost much of its value in the area of knowledge and learning for, repeated by so many thinkers, it had sunk to the level of a platitude. The account given by Rheticus, on the other hand, together with some statements made by Copernicus (as for instance 'the ballet of the stars' of which he writes in his *Commentariolus*), testify to the great astronomer's direct interest in music – or rather in musicology. In highly graphic terms he compares the modelling of the new picture of the world with the musician tuning his instrument.

It has recently become evident that Copernicus was not only endowed with an auditory but also with a visual imagination. To explain this, it is first necessary to remember that the Ptolemaic model of the Universe was based on Aristotelian physics which assumed that the Earth, being the heaviest element, is in the center of the concentric system of spheres. On the other hand, the spheres of water, air, fire and so on, arranged in a concentric order, soared above the Earth in that the Earth, it was alleged, was as an island in an ocean, bound on all sides by water. This system of spheres, as Thomas Goldstein points out, was presented by the traditional cosmographies which continued to appear as late as the 16th century. Notable among these was the *Cosmography* of Apianus, published for the first time in 1524. But at the same time, other maps that began to appear with the beginning of that century represented a different picture of the world. They took account of other continents, including Africa and the just discovered America. It was now more difficult to think of our continent as an 'Earth island' occupying a special position in the hydrous sphere and surrounded by heterogeneous spheres. With this picture of the universe in mind, Copernicus was able to imagine the Earth as a homogeneous spherical heavenly body composed of two fundamental elements (water and land). In support of this thesis he pointed to the new geographical discoveries. The heliocentric model of the Universe, which was also the product of his reflection upon the new map of the world, not only repudiated the Ptolemaic geocentricism but also subverted Aristotelian physics, which made a distinction between the laws that governed the sublunar and the supralunar spheres. When we recognize this fact, we must remember that the premises of Copernican reasoning, closely inter-related and based on an auditory and visual imagination, comprised one 'imaginative' argument which reinforced the 'rational', that is strictly astronomical, evidence and it was this evidence which led to the crystallization of the heliocentric theory.

This was one of the most original arguments devised by Copernicus, the reformer of ancient Pythagorean thought. The remaining arguments were of a different nature for they sprang from the reasoning of a mathematician, that is of a man who plotted the positions of heavenly bodies. No wonder then that his contemporaries thought of him primarily as of a mathematician. Rheticus, who held the chair of mathematics (or rather of mathematically oriented astronomy) at Wittenberg, called Copernicus in the title of *The First Account* a 'remarkable mathematician' and presented himself as a 'youth enamoured of mathematics'. Others too considered Copernicus a mathematician, that is to say a thinker who bestowed upon the heavenly phenomena he observed a mathematical, or the most perfect, form. Traces of the Pythagorean idea may be found here as well, for it was the Pythagoreans who associated mathematics closely with astronomy. Aristotle described their views in the following manner:

> Having become preoccupied with mathematics, they began to believe that its principles are the principles of all existence... they believed that numbers are of primary importance in nature, that roots of numbers are the elements of life, that the entire heaven is harmony and a number.

In turn attention might be drawn to the fact that of all the sciences, mathematics alone enjoyed that honor. Over many centuries it has been considered the peerless ideal of all the other scientific disciplines. Without going as far back as Plato (who, as legend has it, did not permit anyone who had no knowledge of mathematics to approach him), one might mention Roger Bacon, who wrote in the 13th century

> that all science requires the application of mathematics,

or the praise of mathematics contained in *Paragone*, a work written by Leonardo da Vinci three hundred years later, during the lifetime of Copernicus. Leonardo wrote:

> no human reasoning, if it has not passed the test of mathematical reasoning, deserves to be trusted.

In the eyes of scientists and scholars of that and of a later time, mathematics alone was recognized as a true science owing to the certainty it guaranteed to conclusions. Other sciences, on the other hand, were always subjected to crushing criticism. Using the term mathematics today, we tend to think of it in its modern meaning. We think of mathematics as of a discipline whose purpose it is to give numerical order to phenomena, although it must be admitted that this modern interpretation was not altogether foreign to ancient scientists. However, we forget very often that in the 16th and the 17th centuries mathematics was usually identified with mathesis universalis (universal learning), a term which covered a congeries of principles of various branches of science as well as the general concept of method, a fact which many thinkers emphasized. They wrote musical works which were actually mathematical treatises, for the term 'mathematics' included music in addition to arithmetic, geometry, mechanics, optics and hydraulics. The rules of reasoning contained in the Euclidean concept of geometry were usually given first place in this concept of mathematics. Not only a large number of scientists, philosophers and social scientists but also a broad circle of the educated population placed an equation mark between mathematics and the general concept of scientific method. It is thanks to this attitude that mankind can claim very significant advancement in all the branches of learning.

This concept of mathematical reasoning called for the acceptance of a few fundamental axioms which led to the evolution of reasoning, that is to the evolvement of logically constructed conclusions. That is why Copernicus took the working hypothesis as the central thesis of *On the Revolutions*. He had to do this if he did not wish to be buried by the growing avalanche of material gathered in daily observations. We cannot fail to recognize the modern aspects of this style of thinking. It is based, as we can see, on the principle that a conclusion must not be pressed further if in the course of investigation it is discovered to be untrue and that, on the other hand, evidence confirming the validity of a still unproved thesis may be collected in order to enable the investigator to judge at a given moment whether the thesis truly deserves to be called a new theory and if so, whether it should replace an earlier theory that now deserves to be rejected. This is the one and only way of achieving progress in the area of science.

Although Copernicus left no treatise dealing with methods of scientific thinking, yet in his principal work he chose the scientific method of procedure. A closer examination of the line of his reasoning reveals the fact that the book *On the Revolutions* owes its conspicuous position in the history of the evolution of the human intellect to the rare gift for reconciling the exhilarating visions that visit the minds of the greatest scientists with the strict rigors of scientific thinking, and it is the latter that ultimately determine the rank the work occupies among all other works of a given age.

3. Nicolaus Copernicus – Architect of the New Earth

Unlike the works of great artists, even the most remarkable discoveries of scientists age very quickly. To what then does the book *On the Revolutions* owe its phenomenal vitality a full five hundred years since the birth of the man who wrote it? Although made in an age when, far more often than today, scientific discovery was a marvelous intellectual adventure, the discovery of Copernicus still remains a classic example of the power of the human intellect. It might be worth recalling here that astronomy, the most poetic all of sciences, bestows one of the most beautiful gifts upon its followers: it enables them to combine the strict rules of reasoning with the infinite wealth of the imagination. The specific nature of the discipline Copernicus pursued would not have guaranteed this extraordinary vitality to his discovery had it not been for the fact that he invested it with a tremendous charge of disciplined ingeniousness. It is for this reason that the word 'thinker' is a far more suitable description of the nature of Copernicus' mentality than the term astronomer usually applied to him. What is implied here are not his universal interests, but the fact that, observing the Sun, Moon and the stars, he discovered new horizons of human thought to the readers of his work. Copernicus has been justifiably called "the architect of the new heaven"; it ought to be remembered, however, that he also created a new picture of the Earth, or to put it more precisely, that he assigned a new position to the Earth in the Universe and consequently a new position to man on the planet he inhabits.

At the time when Copernicus was sixteen years of age, the Italian humanist Pico della Mirandola wrote in his famous discourse, *On the Dignity of Man*, that the Creator turned to the first man with these words:

I have placed you in the center of the Universe so that you should the more easily see what is around you.

Writing these words he did not anticipate that only a dozen or so years later an idea would emerge north of the Alps which would topple that age-old conviction held by those who lived and died over many centuries believing that they had been destined to live in the center of the Universe. Whenever we say today that the Sun is 'a star in the suburbs of the Galaxy' and that the Earth is a 'space ship' soaring through infinite space, we nearly always remember that the ingenious idea of Copernicus lies at the root of these concepts. Thanks to him, the term 'the planet of people' has been reduced in its significance to a much more modest role than the one ascribed to it by tradition. This has come about because the comparative scale, as applied by man in relation to himself and to the Earth he inhabited, has become altered. Herbert Butterfield quite rightly observed in his book on the *Origins of Modern Science* that when objects were seen in their right proportions Divine effort began to seem too great and the objects of God's exertions too insignificant. That is why the now generally accepted principle of the homogeneous nature of the Universe has been called 'the Copernican principle'. It states that the Earth's position in the Universe is distinguished by no remarkable feature from the position occupied by the vast number of other heavenly bodies.

Unlike Copernicus in his day, we today do not imagine the Universe as finite and fixed in immutable harmony. We know that the Universe is constantly changing, that it may be compared to the ceaseless flow of a stream rather than to a clock constructed once and for all time. Whether this is true or not remains to be seen because the history of science teaches us that the latest discovery is not necessarily true. Yet in the current controversy on the structure of the Universe, the name of Copernicus crops up as frequently as it had at the time when his system roused such fierce passions among the scientists of the 16th and the 17th century. Albert Einstein suggested in *Evolution of Physics*, a book written together with Leopold Infeld, that Copernicus' dispute with the views of Ptolemy was of no significance because in the light of the theory of relativity the choice of one or another system is merely a matter of agreement regarding one of the various coordinate systems. Einstein was told at the time that he was wrong because nearly all of the matter of the solar system is concentrated in the Sun, whose mass is 330 thousand times greater than the mass of the Earth and 750 times greater than the mass of all the planets taken together. The point was also made that we cannot always chose any arbitrary frame of reference. This is not possible especially when we wish to study not only motion but also the rules that characterize it. Thus, for example, when we observe the motion of a mote in a beam of sunlight we can take as our frame of reference any random speck of dust. When, however, we intend to study the laws that govern that motion, we must select one priveleged system. Within the compass of the universe, the modern defenders of the Copernican theory state: the Sun is that privileged frame of reference where the laws of the gravitational field as well as the special theory of relativity are in force. Consequently, the Copernican heliocentric system is not only not superseded by the theory of relativity but can be coordinated with it as a concept defining the choice of the proper frame of reference.

This is yet another argument proving the continued vitality of the Copernican theory both in its astronomical and its general philosophical aspects. It might be brought to mind again that Copernicus bequeathed to posterity one other idea related to a problem which carries great weight in the study of the reality around us and, what is closely allied with it, which has a direct bearing on the picture of the Earth and of the Universe as it is increasingly perfected by man. The great German Romantic poet Johann Wolfgang Goethe, who was also a remarkable scientist, noted that when it comes to testing an idea in the area of natural phenomena, the most misleading fact is that frequently, or rather nearly always, the idea belies our sense perceptions. He went on to give a salient example of this:

> The Copernican system is based on an idea that is difficult to understand and that constantly contradicts our sense perceptions.

This statement supplements the fairly one-sided opinion expressed so often about Copernicus where the accent is placed on his dispute with traditional authority while a no less important subject, one that is crucial to the development of a scientific picture of the Universe, is cast in shadow. That subject is that the theory formulated by Copernicus was in complete disagreement with the picture of reality as perceived by our senses. Copernicus himself was well aware of this. He wrote in the preface addressed to the pope that his work will probably seem 'absurd' and, in Book I.8 of *On the Revolutions*, repeating the ancient argument, he drew attention to the fact that we are merely deluded by our senses because when we sail from a port, the land and the cities seem to recede. To mariners on a ship sailing the calm seas, it is the external objects that seem to be moving backward while they have the impression that they, together with everything on board ship, remain stationary.

So, Copernicus belonged to the group of Renaissance scientist who had grasped the true essence of human knowledge. He conceived the evidence of the senses and the intellect as interacting factors that are indispensable in the discovery of reality. He knew that the human intellect, obtaining impressions through the senses, does not remain passive but that it processes the data transmitted by the senses. 'Experience' and 'intellect' often covered a variety of meanings under one common term and conveyed the same meaning under different terms, while taken together they reflected the humanistic as opposed to the theological attitude: a critical attitude opposed to dogmatic thinking. Because of the authority commanded by traditional dogmatism, experience and the intellect formed a common front against it.

Before scientists had the use of the telescope and the microscope (that is before they could enhance the power of the senses), the human intellect constituted the sole instrument which enabled them to assume a critical view of the things they observed daily. The astronomical instruments used by Copernicus did not differ significantly from those that had been used by ancient astronomers. They could have inevitably led him astray as they had misled so many astronomers who lived before and after him. It is not to be wondered then that he concentrated his efforts on perfecting man's reasoning powers and that he realized, as we do today, that although concepts are indeed born of experience yet sense perceptions are not always transmitted faithfully but that they are always, what the 14th century Italian writer Marsilius of Padua had called, 'rationalized experience'. The human mind selects from a myriad of impressions those that provide data needed in a given research project. Consequently, no observer remains unaffected by earlier findings and by past experience that has already been logically organized. In this manner, every scientific observation of reality is composed of 'sensual' as well as of 'intellectual' factors. Kant and his followers stated that our mind synthesizes 'material supplied by the senses'; later, however, when the discovery was made that this 'material' already contains rational elements, modern philosophers were induced to reconsider the concept of the line dividing 'empiricism' from 'rationalism'.

If it is accepted that Copernicus deserves to be called a thinker rather than an astronomer, then it follows logically that he should be called the architect and not the discoverer of the heliocentric theory. Since to discover suggests a passive attitude toward reality while the term architect signifies a creative attitude, hence one who enriches our knowledge by an active attitude toward the world. However, no matter what we chose to call him, the substance of the Copernican discovery remains unaltered. Regardless of where, among the thousands of millions of heavenly bodies, man has come to live, he is capable by the power of his intellect and imagination, to produce works which serve as roadsigns to many generations of thinkers who pool their effort to develop a new and even more perfect picture of the Earth. The intellectual adventure that Copernicus experienced represents a sensational event which occurs only once in hundreds of years in the history of science. It cannot fail but arouse admiration even in our day and age when so many and such momentous discoveries have been made in the lifetime of one generation.

THE IMPACT OF COPERNICUS ON THE DEVELOPMENT OF THE NATURAL AND THE HUMAN SCIENCES

BOGDAN SUCHODOLSKI

Faculty of the History of Science and Technology of the Polish Academy of Sciences

Erat ad omnia quae non essent philosophia minus attentus

The title of this brief article, whose purpose it is to speak of the influence Nicolaus Copernicus exercised not only on the development of the natural but also on the development of the human sciences, may evoke some surprise. Nearly all of us have been brought up in the ideals of the 19th century, a time when the natural sciences and the humanities came to a parting of the ways. We are the product of the 19th century in the sense that we remember the efforts made to establish separate methodological principles for the two areas of human knowledge. We live in the 20th century, a time when the conflict between the natural sciences and the humanities, expressed by the antithesis between culture and technology, seemed (and still seems) unsolvable to many of us. With the basic values of our culture threatened by the dangers of war and destruction, with certain areas of our life menaced by a socially uncontrolled development of technology, we are ready to believe that the conflict between the natural sciences and the humanities must grow more dramatic.

Reflection on Copernicus and his work, hence reflection on the humanistic aspects of the evolution of natural science doctrines, refutes our experience. Indeed, the most important factor in the reactions to Copernicus and his ideas throughout history is the fact that the natural sciences became the frame of reference of the deep controversies on world outlook (Weltanschauung). Consequently, any reflection upon the work of Copernicus must include not only the history of astronomy but also the history of man in his role as the stargazer. The two aspects, natural and humanistic, are indissolubly bound in the activity of Copernicus and in the posthumous fate of his doctrine.

Formulated in the first half of the 16th century, the new Copernican science of the universe seemed at first to hold interest only for a small group of astronomers. The late 16th century and nearly the whole of the 17th century made it clear, however, that the new science was then considered important by all who wished to gain deeper insights of the condition of man within the compass of the great world of nature. Adherents of Copernican science were persecuted by the Church not because they believed in the principles propounded by that science on the subject of the stars, the Sun and the Earth, but because the astronomical theses gave rise to new ideas regarding the people who inhabited that Earth. The Church was not directly interested in the science of the Earth, Sun and the stars. It was interested in what people thought of their status in the universe. It is in this context that the indissoluble bonds between the natural sciences and the humanities manifested themselves in the work of Copernicus and in its history after his death.

Which experience reflects the truth about man and nature? Is it the experience of the 19th century which sought to persuade us of the inherent contradiction between the humanities and the natural sciences, or is it the historical Copernican experience, which pointed to the links between the two?

We believe that the truth, the great truth of European history, lies in the route blazed by the Copernican idea. Further reflection on Copernicus and his work leads us to believe that the roots of the belief in an alliance between the concepts of nature and concepts of culture lie in the Greek tradition, that they go back to the Greek civilization when the idea of the world and the idea of man were indissolubly bound. It was primarily Greek art, and specifically the Greek drama, that provided evidence of man's close links with the universe. The terms 'logos' and 'ethos', whose meaning was developed by Greek philosophy and Greek drama, defined man as a creature submissive to the great life powers and at the same time as a creature rebelling against the fate that the stars had in store for him. In the Greek tradition destiny reigned supreme over the world of gods and the world of men alike, but man's rebellion against destiny, a rebellion symbolized by the burning torch of Prometheus, was also in the Greek tradition. The burning torch was conceived as a symbol of man's burning passion and it is this passion which enabled him to make such great strides in the field of technology. Through the light of knowledge and through the power of technology man was to gain mastery of the universe of which he was a part and which, though it was hostile to him, could at the same time be obedient to his will.

European culture, social life, metaphysical concepts and art, assumed in the course of evolution forms which differed drastically from those that prevailed in Greece. Yet Greek drama and Greek mythology remained alive in the European consciousness precisely because the problem of man's place in the universe continued to be as important as it had been in the past, and because the advance in the natural sciences and the advance in our knowledge of man were interrelated.

The Renaissance, the age of Copernicus, discovered new dimensions to these problems. In the revolt against medieval concepts of the world and of man, the men of the Renaissance, the great artists and poets, the great travelers and scholars, the great politicians and commanders, discovered a new face of the Universe and of the Earth. They gave a fresh reading of the ruins and of the mementoes of the past, they organized states and cities on new principles and developed a new concept of human happiness.

This was an age of great discoveries which disclosed the new shape of the material world; this was an age of great courage and of equally great risks, of great joys and of human victory over ignorance and submission.

There is something symbolic in the admiration the early Renaissance had for David, who against incredible odds defeated a much more powerful enemy by the sheer courage and strength of his youth. The remarkable 15th century Florentine sculpture by Donatello shows the young victor contemplating his victory. Andrea del Castagno

Portrait of Nicolaus Copernicus by an anonymous painter *circa* 16th century.
(Photo courtesy Magdalena Rusinek)

The Copernican Tower in Frombork.
(Photo courtesy Magdalena Rusinek)

also shows David after the contest as with hand raised high he strides amidst the enthusiasm and adulation, his fingers spread out like claws against the lowering sky announcing the triumph of greatness and the joy of victory.

At the close of the 15th and the beginning of the 16th century, the symbol of triumphant man was portrayed by Giorgione in his arresting picture of the three philosophers. The two older men contemplate the world with sadness and anger, the third, a youth, holding a pair of compasses and a square in his hand, looks out with hope and wonder at the distant landscape, at the heaven and the Sun which he will study.

We shall never know whether the young philosopher was, as legend has it, Copernicus. But even if the third philosopher is not his portrait, it nevertheless remains a symbol used by Giorgone to depict the new, that is to say the Copernican, method of discovering the Universe. It brings out the contrast between the new and the old abstract and verbal methods.

The new method of investigating reality was employed by the great travelers, among whom Columbus was the first, in their bold and fearless expeditions which led to discovery of new lands. It is this method which enabled Machiavelli to lay bare the hidden mechanism of political struggle and of the structure of state authority. Improving upon that method, Leonardo da Vinci and Dürer ventured beyond painting to develop theoretical methods for discovering the material forms of nature and of the human body. The masterpiece of Copernicus on the structure of the universe, published in the same year as the equally remarkable and bold study of Vesalius, *De humana corporis fabrica*, represents the ultimate triumph of that method.

Although a variety of specific recommendations regarding scientific investigation were made, yet trust in the intellect, especially in an intellect which took recourse to mathematics, constituted the substance, the spirit of the new scientific method which inspired boldness in human thought and action.

The greatness of mathematics was often lauded in that age. Copernicus belonged to the fold of its enthusiasts. He stated openly that simple mathematical calculation spoke in favor of his theory. He wrote:

> I think that it is easier to believe this than to confuse the issue by assuming a vast number of Spheres, which those who keep Earth at the center must do. We thus rather follow Nature, who producing nothing vain or superfluous, often prefers to endow one cause with many effects.

That is how Copernicus formulated the major problems of modern epistemology: that is problems relating to the human intellect and to the laws of nature. How then do we explain the fact that truth as conceived by the intellect is invariably the same as the laws governing reality? Does this mean that the workings of the human mind comply with the demands placed upon it by 'the wisdom of nature', or does the wisdom of nature conform to the demands put upon it by the structure of the intellect? It is in this context that Copernicus characterized his theory which gave an ordered mathematical picture of the universe. He said:

> So we find underlying this ordination an admirable symmetry in the universe... such as can be discovered in no other wise.

Analyzing the works of Copernicus and their history after his death from this standpoint, we note a significant maturation of processes in the European civilization which linked progress in the natural sciences with human advance in self-understanding.

Copernicus emerged in Europe in a period marked by the first major crisis in the confidence placed by man in his own powers of understanding.

What was the nature of this first crisis of confidence in the history of Europe? This was the first time that the form of confidence to which the great scholastic system subscribed in the Middle Ages was called in question. The scholastic system was the expression of human confidence in the powers of the intellect and in their specific interrelation with the metaphysical structure of life. Scholasticism was based on the belief that there was a kinship between the human intellect and the divine being. Due to this kinship the success of the efforts to broaden the range of human knowledge was insured by the metaphysics of man's aspiration to perfection.

The Renaissance on the other hand cast doubt on this form of confidence, demonstrating that it is a dogmatic principle which takes no account of reality. The Renaissance disclosed a totally new aspect of confidence in man's powers of discovery, that is in the sensual knowledge of an empirical reality. Knowledge of the natural world and history was gained by direct observation. The Renaissance therefore witnessed the evolution of understanding through sensual perception of concrete events, of the natural environment and geographical communities, as well as of the past, of tradition, historical relics, monuments and ruins. Knowledge and understanding were gained directly through the agency of the senses.

Consequently, the Copernican concept of knowledge, opposed to the two former concepts of confidence in the powers of the intellect, emerged in a climate characterized by a crisis of confidence in dogmatic and scholastic rationalism and by the rise of confidence in the sensual perception of concrete reality. Copernicus blazed an important new path for modern scientific knowledge which was opposed to dogmatic intellectual speculation to the same degree as to the acceptance of data resulting directly from sense perception. The greatness of Copernicus was often attributed to the fact that he had the courage to oppose the ancient authorities and established dogma. However, his greatness might be more largely attributed to his strong opposition to the acceptance of data derived directly from sensual perception as well. He was careful to explain that the Sun neither rises nor sets in the heaven. Consequently, the scientific method of Copernicus was an act of great courage not only because it contradicted ancient, Biblical and scholastic tradition, but it was an act of great intellectual courage also, or perhaps primarily, because it contradicted the data obtained directly from general sensual perception as well.

The new method of understanding evolved from the days of Copernicus, and it is thanks to this method that modern science has made such great progress. But the mathematical perception of the Universe remained in conflict with the picture of the world obtained directly through human experience.

While science described the universe in increasingly abstract terms, 17th century art, particularly Flemish art, described the world as a world of direct sensual experience. From the point of view of this art, the human environment was a world of sensual qualities rather than of mathematical equations.

John Locke, who made such an important contribution to the scientific methods of investigation, noted the dual nature of the two concepts of the world with an incisive clarity. Analyzing our impressions, our concepts, as we would say today, Locke sought to convince himself that, although they represent a subjective human state, they are nevertheless man's authentic reality. Does the color blue truly exist? Locke asked. Is the sky really blue? Are flowers blue in color? There's no point, according to Locke, in asking that question. Since the color blue exists as an innate human experience there can therefore be no doubt that it exists no matter what the natural sciences have to say on the subject of its true extra-human existence. There's no point in asking whether the color blue is true or false because it does in fact exist. Art, according to Locke, should deal with subjective reality and we are not qualified to say whether this reality corresponds to objective reality in any way or not, even though we may know beyond any doubt that it is our human reality. Science, on the other hand, deals with a reality that is not the subjective reality of our experience. It deals with a reality which represents the truth about life, informing us what it really is.

That is how the singular dualism of the two different kinds of human confidence in man himself became established in the 17th century. In the first instance, man trusted his sensual perception of the reality in which he lived and which was presented to him by art; in the second case, man placed his trust in the non-sensual and non-conceptual that is in the mathematical and abstract understanding of reality which, though it faithfully reflected reality itself, was not faithful with regard to human experience and human perception.

Returning to the problems of our day, we could ask in this connection: Has this dualism continued into our age and will this dualism continue into the future? Or is it indeed, as some modern scientists and philosophers would have it, that we should accept the thesis of two human cultures: the scientific, mathematical, abstract, invisible and inconcrete culture which is in no way connected with human imagination and emotion, or with concrete human experience, and the second – the pictorial, plastic, concrete and living culture, related to human experience through literature and art?

Upon consideration of the thesis of the two cultures, I believe that we might expect a period of new integration of the natural sciences and the humanities, hence an integration of the two cultures. Another reason why I believe that the integration of the two cultures is possible is because the natural sciences and the humanities are today emerging in a new form, one that bears no resemblance to the forms which led the two into conflict in the 19th century.

This new emerging form of the natural sciences and the humanities is connected with the conviction that truth is attained not so much by the simple portrayal or miming of reality as by the construction, by the creation of reality. The natural sciences are now emerging from a stage of development when the only concern was observation of reality and are entering the stage of development where observation of reality must be related to a creation of the reality which is the object of observation. In nature atomic energy remains concealed until it is released by physicists, only then can it become the subject of study and observation. The same is true of the social sciences. It is not enough to observe the course of history in which man has manifested himself in order to understand the nature of man. It is necessary to create a new social reality in which the true nature of man, as he is and as he can be, is revealed.

The conviction that we acquire an understanding of reality by creating a new reality is now the thesis put forth jointly by science, technology and art. The conviction that truth is the expression of human creative powers which give new shape to reality is the conviction which provides a basis on which the new integration of the two cultures – the scientific culture and the art culture – can occur. One may of course take the catastrophic view that science and technology will ultimately destroy the natural environment and human life. But there seem to be more grounds to the belief that a broader knowledge of the universe will become the source of a new power which will enable man to construct a reality that will measure up to his dreams and expectations.

And when we look back from this vantage point upon the work of Copernicus, we note how much we have in common with him. Yet when we remember that his scientific investigations were conducted in the lonely tower of Frombork, practically with no help of instruments, and when we compare it with the present situation when man can reach the stars not only through extremely precise giant telescopes but also through flights in space, we then see how far we have come since that simple and valiant age. However, we also see that the problems of that distant age are still our problems.

Copernicus convinced that the intellectual life of the individual who gains knowledge of the universe by new methods must differ in its form from the intellectual life of persons who obediently accepted the traditional picture of reality. He convinced that the new type of intellectual culture calls for new intellectual forms of self-knowledge. Since in our atomic age knowledge of nature is expanding another degree, human life grows correspondingly more complex and is weighed down by greater responsibilities. And as in Copernicus' time so now too we are concerned with the laws which enable us to gain greater knowledge of the heaven, and as before we are now seeking the ideal of man's greatness, we wish to establish the position of the Ea·th in the Universe and man's position on Earth, hence as in the past we are concerned with the grand alliance of the natural scientists and the humanists.

The growing confidence in the powers of man's understanding, a distinctive feature of the self-knowledge of Renaissance man, was connected in a most dramatic manner with a growing sense of concern aroused by the exposure and destruction of the traditional picture of the universe and of man. The boldness of human thought and action undermined the order of the medieval cosmos where man, his life, tasks and responsibilities all had their appointed place.

Did this boldness launch man upon the path to happiness? This perplexing question was posed with growing frequency in the late Renaissance period.

In Italy Savonarola, who denounced the Florence of the Medicis, the ostentatious way of life and the splendors of its culture, became the symbol of the wane of the Quattrocento. Ficino cast his 'pagan' compositions into the fire, Botticelli burned his pagan pictures. What then is the true meaning of life? What is the true substance of human creativity? Does wealth insure happiness? Carpaccio's unusual painting at the Museo Civico in Venice depicts two wealthy Venetian women sitting on the terrace of a house in the company of a dwarf, a peacock and curiously shaped dogs. The women are gazing into space in numb and mindless expectation of a miracle which might relieve the utter boredom and inanity.

In the meantime, social and religious conflict grew in violence north of the Alps. The secular powers and the spiritual powers were losing their influence. Peasant wars, particularly the revolt led by Thomas Münzer, though it was savagely suppressed, inflicted a blow to the secular order of life. Although Luther defended the prevailing social order and though he demanded that the rebellious peasants be punished, he nevertheless inflicted a crippling blow to the religious order of the Roman Catholic Church.

Cruelty and destruction were rampant in the climate of insecurity. Dürer depicted the age in 15 great woodcuts, which he entitled the Apocalypse, and presented himself as the vir dolorum. Bosch in his pictures portrayed the boundless cruelty of life and the deceptive enticements and delights of hell that hold sway on earth. In this terror-filled world, a 'true man' like Christ must carry his cross amidst the jeers of the sages and the rabble. This world is seen as a 'ship of fools' in the popular version of protest against it. The great humanist Erasmus of Rotterdam was to describe the unchecked madness in his tragic and ironic reflections on folly. His friend Thomas More of England lost his head for exposing a social order in which the 'sheep devour the people'.

As the traditional order of life crumbled and as at the same time Renaissance hopes were blighted and the brilliance of the Renaissance dimmed, a lone Polish astronomer was about to deliver another blow to human arrogance and human confidence. The earth on which the human kingdom was proving itself so inhuman was not to remain the motionless and central point of the universe but, pale and alone, it was to revolve as one of the stars in a clearly described orbit.

Thus the power of human understanding, which had at one time generated a sense of greatness in man, now revealed a structure of the universe that became the source of fear because it made man aware of his puny stature and solitary condition. The Universe was no longer a cosmos created by God as the home of man; the order of the universe was becoming a mathematical order which, though greater in terms of human thought and emotion, was alien and cold in terms of human experience.

Only a small number of mathematicians were able to evaluate Copernicus' theory as a triumph of the power of human thought, equal in importance to the theory of the mathematical harmony of the universe. For many centuries the theory evoked indignation and fear in the broad circles of the population which included scientists and scholars. The concept seemed to diminish man's importance and to destroy his exalted position in the universe. Even men of such stature as Melanchthon, Bacon and Komensky could not forgive Copernicus. However, no one was as profoundly affected by the dramatic contradiction inherent in the greatness and the insignificance of man as was Pascal. Although he worshiped mathematics, he was also deeply responsive to the needs of the human condition. It was Pascal who realized that from now on man's greatness and his insignificance in that new Universe, his power and his ineffectiveness were to be his inexorable fate.

Was Copernicus aware of the dramatic contradiction in which his science of the earth and the stars had embroiled the human race? There is much indication that Copernicus was fully aware of the importance of the theory he proposed, an awareness that was attended by an apprehension that his concept might be ridiculed and proclaimed absurd or perhaps even blasphemous. It is known that he hesitated a long time before he decided to allow the results of his studies to be published.

There was no indication, however, that Copernicus did in fact experience the reactions which his criticism and refutation of the traditional concept, saying that the Earth is the motionless and central point of the Universe, aroused among broader and broader circles as his ideas spread throughout Europe.

The writings of Copernicus, and most likely his consciousness, were pervaded by an atmosphere of serenity and harmony, by a sense of dignity and joy. There is no trace in them of a sense of impending calamity and no presentiment of the storm that would break over this theory.

Clearly, Copernicus could not have foreseen that a half century later Giordano Bruno, an enthusiast of his theories, would die at the Campo di Fiore, or that some years later another of his disciples Galileo would fall into the clutches of the Inquisition. Even so, it is astonishing that he should not have realized that the truth he proclaimed would shock rather than bring comfort to people.

How do we explain this? Why was it that he, who for the next two centuries at least was to become the cause of human anxieties as well as of violent polemics and cruel persecution, was, if the historical evidence of the Renaissance is to be believed, a universal man, a man marked by an inner harmony which was in consonance with the great harmonia mundi?

I believe that the answer to the question will become clear once it is duly realized that only one aspect of the Copernican concept impressed itself upon popular consciousness. It was generally believed that the theory of Copernicus had dislodged the Earth from its privileged position and set it in motion. That is how the theories of Copernicus always were and still are interpreted. But this accounted for only one aspect of his concept. It ought to be remembered that Copernicus not only 'moved the Earth', but that above all 'he had stopped the Sun'. His teachings presented a grand concept of the universe whose center and life-giving force was the Sun. That explains why his theories were so harmonious and joyful.

Copernicus wrote:

In the middle of all sits the Sun enthroned. In this most beautiful temple could we place this luminary in any better position from which he can illuminate the whole at once? He is rightly called the Lamp (the 'lucern amundi'), the Mind, the Ruler of the Universe. Hermes Trismegistos names him the Visible God, Sophocles' Electra calls him the All-seeing. So the Sun sits as upon a royal throne ruling his children the planets which circle round him. The Earth has the Moon at her Service. As Aristotle says, in his *De animalibus*, the Moon has the closest relationship with the Earth. Meanwhile the Earth conceives by the Sun, and becomes pregnant with an annual rebirth.

Copernicus was not alone in his heliocentric views. The Renaissance had a great enthusiasm for the Sun. A combination of ideas contributed to this enthusiasm: Neoplatonism and the Pythagorean traditions, Stoicism and some beliefs incorporated from religious mysticism and the Franciscan ideals, medieval study of optics connected with the philosophically important distinction between *lumen* and *lux*. We might recall in this connection the numerous writings of Ficino, especially his *Orphica Comparatio Solis ad Deum* (1480) in which he described the Sun as a life-giving force, as the cosmic tabernacle with its visible and imagined creative power and as the king sitting on a throne high above the planets he governs. We might also recall how fascinated Columbus was with the Sun.

Need we be reminded of Leonardo da Vinci's notes on the subject? We read there that the Sun is immobile in the Universe and that it sheds its light on all the heavenly bodies found in the Universe, that all the souls come from it because the warmth that dwells in living creatures comes from the soul and there is no other warmth or light in the universe but the Sun's. With this concept of the Sun in mind, Leonardo spoke with indignation and anger about those who subscribed to the idolatry of man rather than of the Sun, seeing no other more magnificent and more virtuous body in the Universe.

The Renaissance concept of the Sun is reflected in the theories of Copernicus. The Universe as conceived by Copernicus was finite where the Sun, the great 'lamp of the world', the magnificent tabernacle of divine providence, was king. The Earth was said to revolve under the Sun's invigorating rays serene, safe and happy thanks to the light and warmth of the Sun.

It is worth remembering that in spite of the general opinion held about the Copernican system by the religious people – catholics and protestants – it is this sense of hope and joy that Cardinal de Bérulle perceived in it. At a time when the Inquisition prosecuted Galileo, he stated that

> un excellent esprit de ce siécle a voulu maintenir que le Soleil est au suivie en la science des Astres, est utile, et doit être suivie en la science de salut. Car Jésus est le Soleil immobile en sa grandeur, et mouvant toutes choses. (*Discourse de l'Estat et des Grandeurs de Jésus* – second discourse, 1622.)

Although Cardinal de Bérulle was fairly isolated in his judgment of the metaphysical and religious significance of Copernicus' theory, the idea of the 'solar universe' was winning far wider importance in Renaissance Italy as witnessed by the proliferation of architectural and social ideas regarding the construction of new and happy cities. This aspect of the idea was finally expressed by the works of Campanella.

Campanella presented his vision of social order and beauty in a work entitled, in difference to the rich ancient and Italian tradition, *The City of the Sun*. The work may be taken as a philosophical counterpart of Copernicus' cosmic vision of the solar universe.

But we must not forget that *The City of the Sun*, described by Campanella was a

Utopia. His vision was a dream or at best a hope. It was not a realistic vision. Did that mean that man does not belong to the solar world? Did that mean that man is bound to the cold Earth?

Let us return for a moment to the heliocentric concept of the early Renaissance. Ficino's enthusiasm for the Sun was not shared by everybody. Ronsard, the French poet of joy and love, although he did indeed follow the path indicated by Ficino in his cosmogenic poems – *Hymne des Daimons*, *Hymne des Astres*, *Hymne des Étoiles* – and in his *Discours de l'excellence de l'Esprit Humain*, indicated nevertheless that Saturn and his malencholy is the antithesis of Apollo, that the stars, the daughters of night, are the antithesis of the Sun, the 'flaming and glowing soul of the world'.

Leonardo, though he worshipped the Sun, also knew that man cannot reach its greatness even with his eyes. It is in this sense that he spoke of the *splendore del corpo solare nelquale non si puo fermare l'occhio umano.*

Michelangelo presented this dualism in the most dramatic manner. His Last Judgment in the Sistine Chapel, painted when Copernicus was writing his *Commentariolus*, shows the antagonistic forces of light and darkness. Although the interpretations which say that the great fresco is a pictorial vision of heliocentrism might go a bit too far, there is no doubt, however, that Michelangelo's Christ combines the traits of Jupiter and Apollo, that he is presented as the Sun, the giver of life, which at the same time burns everything that lives.

The dualistic concept of light and darkness, depicted with such mastery by him in the allegories on the tombs of the Medicis, was also the subject of some of Michelangelo's verses, notably the sonnet *Ogni van chiuso.* Everything lives and blossoms in the Sun, only man alone grows in the shadow: *Ma l'ombra sol a piantar l'uomo serva.* Only man, who sees the stars, values the night more than the day.

The thoughts expressed by Michelangelo in his paintings, sculptures and poems are the subject of the theoretical works of G. Bruno and B. Telesio. The fifth dialogue of Bruno's *Eroici* in particular, referring to the myth of Artemis and Actaen and to the anger of Apollo, demonstrated the internal antagonism between the Sun's life-giving and destructive powers. And although man may be 'Heliotrope', as one of the characters in *Della causa, principio e uno* is called, he cannot look at the Sun. He who looks at the Sun sees darkness.

Bernardo Telesio in his *De rerum natura* (1586) also describes the Sun as the *dato vitae*, the giver of life. At the same time, however, he accepts two antagonistic elements as the structure of life, hence light, brightness, warmth, motion on the one hand and the 'earthly elements' of darkness, immobility and cold on the other.

The life and work of Campanella reflected the dramatic antagonism of the Sun and the Earth. Though suffering untold anguish and destitution in prison, he wrote a hymn to the Sun. Al Sole which is the *padre di naturo* and *degli astri* and he beseeches God to enlighten him, for people have taken the light of the Sun from him – *Venga la luna tua.* He dreams that the future city of people will be a city of the Sun.

But in what sense and in what area does man belong to the city of the Sun, in what area and in what sense is the darkness of the Earth his kingdom?

Renaissance writers saw the dialectic of human destiny in the fact that a life in the Sun was the unfulfilled dream of man's desire. Living in darkness and the cold, man yearns for the Sun, but when he looks up at the Sun, he would perish in its fiery heat. Man, as Michelangelo remarked, is not a pheonix who dies in the Sun and rises again thanks to the Sun.

Only when we accept that Copernicus stopped the Sun rather than that he moved the Earth, can we understand the fate that lay in store for his theory in later history. In the annals of mankind, he was seen above all as the one who had dislodged the Earth from its privileged position and told it to revolve in the universe as one of the stars. The controversy focused on the issue of what Copernicus had done with the Earth and not on the problem of what he had done with the Sun.

What was the reason for this? The reason was that the Copernican concept of the Sun was rather a philosophical and poetic vision unfolded above and beyond strict mathematical considerations. The reason was also that the concept postulated the existence of a finite universe in which the Sun was believed to reign. It was this concept, a reversion to ancient and medieval tradition, that could not be sustained for long.

Rarely in the human annals has such a violent conflict been witnessed between the greatness of a man and his work and the historical significance of what he had accomplished as in the case of Copernicus. For many centuries, even to this very day, he remains in the minds of men as the one who had moved the Earth. For this one fact he was deprecated by some and worshipped by others. He, however, was convinced that the system he created had stopped the Sun and placed it at the center of the Universe. Yet it is this idea that was displaced shortly after the death of Copernicus when, what a tragic irony of fate, his disciples who continued the studies he had begun arrived at the conclusion, as G. Bruno did, that the Universe is infinite. In this infinite universe, the Sun was no longer the center of life and everything dissolved in the infinite cosmic space.

That is why the Copernican idea of the motion of the earth was linked in the later development of that concept with the idea of the infinity of the Universe rather than with the vision of a motionless Sun placed in the center of a finite universe. Thus the Earth was dislodged from its privileged position and set in motion while the Sun did not become the lamp of the universe shining at its center.

More than four centuries have elapsed since the day when the Polish astronomer observed the mysterious starstudded sky from his tower in Frombork. The instruments he used were primitive, the weight of past authorities and tradition hung like a cloud obscuring the picture of the firmament. The truth he arrived at in these difficult circumstances evoked fear and anxiety. His discovery was, as he confessed, difficult and incredible for it contradicted generally accepted views.

In our age of giant telescopes and manned space flights, how deeply we penetrate the infinite and mysterious Universe, how greatly human life has changed on Earth, how intensive and universal the search for the city of the Sun here on Earth has become.

But, though the transformations are so immense and though such a great distance separates us from the man who had moved the Earth, we ask with the same anxiety and hope as he did in his day, what is the true sun of human life? And we still do not know whether man is a creature born in the Sun and living toward the Sun or whether the human realm is the darkness and the cold night which enables man to see the stars.

COPERNICUS AS ECONOMIST*

EDWARD LIPIŃSKI
University of Warsaw

Erat administrator episcopae. Vita incolumi...

In Europe the early 16th century was marked by a rapid rise of capitalism within the bosom of the feudal system of production. Productivity grew, the capitalist type of ownership of the means of production flourished and became consolidated while some of the small producers in the towns and even in the villages transformed into a proletariat. The natural economy disintegrated apace, products became commodities, the money economy expanded vigorously, the domestic market gained importance and the last vestiges of the old economic isolation of towns, regions and districts vanished. The substantial increase in internal and foreign trade was directly related to the growth of productivity, to the development of the social division of labor, to the erosion of the feudal system of village industries and to an increase in the volume of marketable agricultural produce. Production became the production of commodities, that means that the goods produced become articles of commerce sold on the market.

The world market expanded and grew in volume. The rise in capitalist production led in effect to the expansion of international commerce. The commodities traded on the world market were not only the spices of India, cloth, skins, furs and so on, that is luxury articles, there was also a growing expansion in the trade of grain, lumber and iron. The money accumulated by merchants became capital. In spite of the growth of the internal market, the bulk of the trade was conducted with foreign markets. This fact explains why the economic thought of the time placed so much emphasis on money and foreign trade. Bullionism, that is what Marx called the monetary system (Monetarsystem), developed. The first ideas of 'national economy', a category unknown in medieval times, took shape. The first concept of the wealth of the country was formulated and the ideas of national economy and national economic policies emerged. The conviction took root that a country can be said to be wealthy when it possesses large quantities of gold and silver, and gold and silver could be obtained for commodities exported to foreign markets. Gold and silver denoted wealth because they represented a stable, lasting and universal form of abundance, a surplus and an excess; in other words since gold and silver were not consumed they represented wealth. Gold and silver consequently came to be considered the material embodiments of what at the time began to acquire the greatest importance, that is accumulation of capital and savings.

That is why the principle of 'good money' which originated with the expansion of

* The essay appeared in Polish in *Mikolaj Kopernik. Szkice monograficzne* (Nicolaus Copernicus. Monographic Studies). Warsaw 1965.

international commerce, acquired the status of a crucial factor in economics. The fiscal practices of the Middle Ages now became an obstacle to the growth of productive power. New economic theories were born of that conflict.

It is not at all surprising that Copernicus should become the theorist of 'good' money and that he should oppose the fiscal-feudal approach to money. He analyzed the question on purely economic grounds, that is as a problem governed by objective laws. In the 16th century the Baltic commerce was of far greater consequence to Europe than trade with the new colonies. Furthermore, the Baltic was an immensely important international trade route. Trade in raw materials which were in great demand on the world market, these supplied principally by Poland, was concentrated in Gdańsk. This may explain why the 'modern' theory of money originated in Poland. For there is no doubt whatever that the treatise by Copernicus was the first economic thesis to herald the coming of a new age, that it was the first thesis to venture beyond the confines of medieval and feudal thought.

In his recommendations of good money Copernicus stressed that with good money in circulation the Polish Republic would achieve greater wealth even if, owing to the monetary reform, individuals should suffer temporary losses. The memorial by Copernicus was a declaration of war on the old worn-out concepts of the past era. The ideas of Copernicus on monetary reform did not become popular or win recognition because they took a categorical stand against the abuse of the right to strike new coins for fiscal and private purposes. Although it had been known for a number of centuries that money should have a stable purchasing power and that the minting of new coins should not be treated as a source of royal revenue, the prevailing practice was entirely contrary to these principles. It was the common practice of the majority of countries to debase money, that is to decrease the silver content of coins while maintaining the same nominal value. Rulers usually considered fiscal needs more important than the needs of trade and exchange.

Different kinds of profit were realized from the minting of money. The chief underlying reason was the fact that the nominal value of money was greater than the intrinsic value of the metal. In the 15th century, for example, the profits of the mint of Vienna were equal to 13 percent of the value of the metal of which 7.9 percent represented the actual costs of minting, about 5 percent the profits of persons connected with the mint while only 0.4 percent went to the ruler.*

The master of the mint and the officials drew the biggest profits from the issue of coinage. With the primitive techniques used in minting effective control was impossible. The Polish ordinance of 1526 provided for compensations to the master of the mint to cover the losses incurred in melting down old coins, for there was a substantial loss when old coins, silver objects and waste were melted down (the silver had to be cleansed of impurities). The treasury also paid for the tools. Besides this the treasurer received 10 grivnas per week, the master of the mint was paid 8 grivnas in salary and 2 zlotys for rent as well as 2 grosz for each minted grivna; the two 'scribes' received

* L. von Ebengreuth: *Allgemeine Münzkunde und Geldgeschichte des Mittelalters und der neueren Zeit*, Munich 1904, p. 217.

1 zloty and 12 grosz per week, the archbishop of Gniezno and the governor of Cracow obtained 25 grivnas each quarter year for their part in assaying coins. The treasurer, the master of the mint and other privileged persons made fortunes on the minting of money. Sombart describes the private fortunes built on minting.*

The production of coins was an extremely costly procedure at the time. The refining of silver was costly in itself owing to the high price of the ingredients used in the alloys and to the high cost of melting down old coins and silver vessels where a substantial loss of material was recorded. Each coin was stamped individually with the die.**

According to Ludwik Jodok Dietz or Decjusz (Decius) the cost of minting ('for labor and other costs') was 24 grosz, or 8 percent, for every 10 zlotys. The total difference between the intrinsic value of the metal and the nominal value of the coins (Schlagschatz or brassage) was 30 percent. The greatest profits, however, were made when old coins were 'revoked', that is when they were withdrawn from circulation. They could then be purchased at a low rate. In some respects this was tantamount to the confiscation of the money capital of the citizens. The debasing of money practiced by French kings, beginning with Philip IV the Fair, is well known. The principle accepted in France was that the value of the livre is stable regardless of its fineness. Acting on the principle that 'pecunia vis est et non materia' (money is power and not matter) and therefore the power to release from obligations lay with the state, money regulations then in force provided that contracts could not be drawn up in units of weight.

As the 'nominal theory of money' spread, together with the practice of constant monetary manipulations, consisting of the debasement or 'improvement' of money in order to insure revenues for the ruler (or for the group of feudal lords who profited by the prince's privilege to mint and to change money), it represented an obstacle to productivity growth. That is why the opposition that rose against royal prerogative (dominialism), that is treatment of the minting of currency as a source of royal revenue, as well as against the principle of nominalism was a progressive activity, for it answered the need to create auspicious conditions for the growth of productive powers.

Copernicus had no equals among the authors who wrote on the science of money as far as realism and precision of thinking were concerned. He deserves to be recognized as the first economist to give a precise definition of the law of money circulation, known as Gresham's law (from the name of a much later writer who, incidentally, never formulated that kind of law). The law states that when metal coins of the same name, that is of the same nominal value but of different fineness, are found in circulation then the better and heavier coins containing more silver disappear from circulation. They are melted down, exported or hoarded. Copernicus is also considered the originator of what is called the quantity of money. According to that theory the purchasing power of money is determined by the relation of the quantity of money in circulation to the volume of goods on the market. Copernicus was undoubtedly one

* W. Sombart: *Der moderne Kapitalismus*, 1925, Vol. I, p. 652 ff.

** L. von Ebengreuth: *op. cit.*, pp. 221–222.

of the first to draw attention to the damaging effects of an excessive supply of money, although his idea had no connection with 'the quantity theory' in the strictest sense of the term.

The question of the quantity of money was not discussed in his time. The problem of increasing the quantity of money in order to invigorate the economy did not exist as yet. The cause of the inflationary rise in prices was rooted at the time in the debasement of the currency. The only question to which attention was given in the 15th and and in the early 16th century was that of the fineness of the coins, of the intrinsic value of money as distinguished from its nominal value, valor impositus, that is of its true value, and the question of the value of metal and of the value embossed on the coin.

The 'dominial' theory of money (royal prerogative) was the reigning principle of the Middle Ages. The sovereign had the exclusive right to issue coins just as he enjoyed the exclusive right of establishing weights and measures. The scope of state function was greatly restricted in the Middle Ages because the role of the state was largely filled by the feudal lords. There was no highly developed tax system at the time. The ruling principle was that the sovereign was entitled to tax his subjects only in exceptional cases such as war. The greatest tax was borne by the peasants and the towns while the Church and the feudal lords were exempt from paying taxes. The main sources of the sovereign's income were: royal domains, customs, the coinage (and the debasement) of currency.

With the growth of commerce, especially commerce with foreign countries, with the increase in the production of goods in the 13th and 14th centuries, the conflict between the fiscal interests of the ruler and the needs of trade became apparent. Merchants who bought and sold foreign goods were interested in good money, for only good money, evaluated on the basis of the pure metal it contained, was accepted in international trade deals and at the large fairs.

In view of these facts, opposition grew against the feudal law of minting coins for fiscal purposes while the then general practice of debasing money met with severe criticism. By the late Middle Ages, money became the major economic issue.

The growing importance of the exchange of commodities forced to the fore the interests of those social groups which took part in that exchange, that is of those who made a living and gained profit from trade. Many powerful merchants were also moneylenders. Consequently the debasement of money affected them as creditors. The interests of merchants were therefore in conflict with the fiscal interests of the sovereign.

But the debasement of money was not the only monetary problem. The practice of trimming coins, forgery and speculation (withdrawal of money from circulation), the melting down and the export of money complicated monetary relations, obstructed the expansion of trade and heightened the risks and the insecurity. Confusion in economic relations was produced not only by an excess of debased money but also by the frequent shortages of money and of metal available for coinage. The forces at play were the changes evoked not only by the fluctuation in the production of gold and silver but also by a favorable trade balance which in consequence meant a surplus of metals and a sufficient supply of money. An unfavorable balance of trade, on the other hand, created a relatively short supply of currency which in effect led to complaints about lack of money. The issue of large quantities of debased currency led to a similar development. Two kinds of currency were in circulation, the good money and the new coins. The same process took place: the better money disappeared from circulation and was melted down leaving only the bad money. Again the merchants would complain about the shortage of good money which they needed for their business transactions.

In view of this situation, injunctions against the export of metals and coins were issued at times, merchants were forbidden to travel abroad while foreign merchants were encouraged to visit the country. These prohibitions constituted the basis of the bullion policy.

Occasional large foreign payments generated a greater demand for silver. The market price of silver soared, the better coins disappeared at a faster rate and good coinage was exported in larger amounts. When the mint price of silver remained stable and when it was lower than the market price, the merchants then did not supply silver to the mint, which often meant that the mints remained idle for long periods at a time. Injunctions against the export of metals multipied but the prohibitions quite obviously were to no avail.

In these circumstances the only remedy was to mint debased coins of less weight and reduced fineness. This kind of debasement of money was not evoked by fiscal considerations alone but signified a necessary devaluation, the lowering of the value of domestic currency, which often stemmed the flow of coins abroad and led to the rise in the value of foreign currency. Hence also to a rise in the price of foreign goods.

It seems that in the early 16th century Poland's foreign trade showed a favorable balance. Some products, such as cattle from Moldavia and Hungarian copper, were imported principally for re-export. That Poland had a favorable balance of trade can be deduced from the fact that English and Dutch trade with the Baltic countries was unfavorable for them.* Poland exported grain, timber, cattle, tar and so on, while her chief import was cloth. It is interesting to note that the Cracow mint operated only 16 years in all between 1502 and 1548. The same is true of the mints of Gdańsk, Toruń and of other towns. This fact might be ascribed to the favorable balance of trade and to the large influx of foreign currency.

Although Copernicus studied the monetary system in detail, he did not consider money as the only factor of the economic and cultural development of the country.

Bad money, Copernicus reasoned, ruins international trade, good money leads to prosperity. Foreign trade, however, is not the sole basis of the country's wealth. Copernicus' reflections make it clear that he believed that foreign trade was based on the growth of production, on the increase in the quantity of productive work. Money reform, the eradication of monetary abuses, the guarantee of good currency, the reduction of insecurity, risk and speculation, all these factors created favorable

* See H. Wilson: 'Treasure and Trade Balances: The Mercantilist Problem', *The Economic History Review* (1949), no. 1, p. 153 ff.

conditions for productivity growth, they served to inspire greater diligence and to raise the numbers of the working force, while culture and science flourished in these circumstances. 'Countries where good money is in circulation', Copernicus said, 'possess works of art, first rate craftsmen and an abundance of everything'. Bad money, on the other hand, produced indolence, sloth and permanent idleness. The arts suffered a decline and there was a scarcity of everything.

These formulations give not only a definition of a 'wealthy country', but also present a complete economical theory. In short they represent an effort at an analysis of the way in which a country becomes 'wealthy'. What country, according to Copernicus, can be defined as 'wealthy'? A country which possesses works of art, first rate craftsmen, a country that is free of beggars and lazy people, a country that is inhabited by an industrious population and that abounds in goods. Science and the fine arts flourish in such a country.

We know that during the Renaissance, production, that also means production of a capitalistic nature, was mainly concerned with the manufacture of luxury articles for the Church, for the manors and for the wealthy people. The objects were of a fine artistic quality, they were expensive and beautiful. Apart from the fine costly cloth, other goods such as expensive silk products, jewelry and works of art were imported from Italy, Flanders and England during the Renaissance. The backward countries exported wool, grain and timber, that is raw materials, food and building materials. Enlarging upon his postulate of healthy money Copernicus envisions an ideal country living in prosperity and desires that his native country should approach that ideal. As in the 17th century Holland was the model country so in the 16th century the cities of northern Italy, principally Venice, were held up as a paragon.

Copernicus considered good money as the basic condition of a country's wealth and not as its wealth. Good money was the basis of the import of foreign goods and of foreign trade because, as Copernicus said, foreign commodities can be purchased for good and not for copper money.

Copernicus placed emphasis on the import and not on the export of goods, expressing the specific interests of Poland's port towns whose wealth was derived from their role as middlemen. Copernicus failed to see any connection between exports and imports. In the polemic on the restoration of money his argument was that a country allegedly needs a supply of good currency in order to exchange it for foreign goods. However, even if the mints of Prussia (a part of Poland) issued nothing but copper coins, imports would not have suffered so long as the existing exports, paid for in gold, made it possible for the merchants to pay for the imported goods. True, the debasement of coinage by a country was a factor in the slowdown of trade, this in turn gave rise to speculation and increased risks, but the ability to mint good coins was the result of international trade and not its basis or its cause.

"Experience is the mistress of life," Copernicus said, "she teaches us that countries with good coinage flourish while countries that use base coins fall and die." For bad money leads to "indolence and persistent idleness".

The satement should be interpreted as follows: bad and debased money, by destroying and by placing an excessive burden on certain social groups, specifically on land owners and the poorest masses, and by ruining trade, kills the spirit of enterprise and diligence in the people. This in turn leads to 'a scarcity of everything' as productivity falls and the pursuit of fine arts and of the sciences is neglected. Only countries with good money possess works of art and have large numbers of first rate craftsman and only these countries have enough of everything.

The above formulations are significant primarily because they document how the great scientist imagined an ideal economic structure of society. The social concepts of Copernicus were the direct product of the social ideas of the Renaissance. The 15th and the 16th centuries witnessed the emergence of the industrial age, a new era in the history of mankind. The leading towns of northern Italy – Venice, Genoa and Florence – thrived on industry and trade and not on agriculture. The handcrafts prospered in Venice, Florence and in many other Italian towns. Luxuries, such as cloth, brocades, glass and gold articles, paper, books, works of art and others, were produced there. The arts were pursued and learning flourished in these cities for the benefit of the wealthy people of the Renaissance. Their aristocratic ideal was described by Castiglione, a contemporary of Copernicus, in *The Courtier* (*Il Cortegiano*, published in 1514). The economic structure of Italian towns comprised the basis for the Copernican vision of an ideal society.

Copernicus could not yet have known the concept of 'national wealth', he did however use the term 'abundance of goods'. The term 'abundance of goods' (abundanta rerum) was known to early scholastic thought. Thomas Aquinas was perhaps the first to speak about it. Society, civitas, hence the city, should, according to Aquinas, own everything within its territories that is necessary to satisfy the needs of its inhabitants and should not depend on merchants to import these goods. In the city itself, trade should be conducted directly between the producer and the consumer while the goods supplied from the outside by the merchants were admissable only in exceptional cases, during a shortage of indigenous products. External trade was rather harmful, the town should be self-sufficient; in times of war an autarchy insured a steady supply of goods; trade, on the other hand, led to effeminacy, destroyed the customs of the home and infected with immorality.* Although trade was indispensable, it should not be allowed to expand excessively. The aim of an urban society was 'an abundance of goods.' There was a division of labor in the town, one served another and all lived together in prosperity.**

The economic basis of the medieval town was narrow, productivity was at a low ebb of development, the volume of commodities insignificant. Scarcity of goods was the prevailing rule. It was in these circumstances that the idea of an abundance of goods was conceived. The town should see to it that sufficient quantities of food are available to the population and craftsmen should supply the necessary raw materials.

In Copernicus 'the abundance of goods' had a different historical meaning. 'Abun-

* *De regimine principum*, part II, Chapter 3 (attributed to Thomas of Aquinus).

** *Schreiber die Volkswirtschaftlichen Anschauungen der Scholastic seit Thomas v. Aquin*, Jena 1913, p. 22, Footnote 2.

dance' no longer signified an adequate supply of the basic necessities of life as it had for the population of medieval towns. Abundance now meant affluence, a certain surplus not of food and raw materials, but rather of luxury items which satisfied the needs of wealthy people, of the manors, and of the town patriciate. The increase in the volume of goods was due to greater productivity while food and raw materials increased in quantity due to the introduction of improved farming methods. Copernicus expressed the ideology of the new era, of the age of the Renaissance. When he said that wherever bad money is found there the pursuit of the arts and learning and all else is neglected and privation can be observed 'due to sloth', indolence and idleness, he went beyond 'the range of medieval thought'. The idea of combating sloth, indolence and idleness was familiar to medieval thought, that is true, but when Thomas Aquinas condemned sloth and indolence, he cemented and provided moral support to the existing system of social organization, privileges and duties. The peasant and the retainer must be diligent and obedient to their master, sloth is a sin. Copernicus, on the other hand, condemned indolence and idleness not because he wished the working man to perform his duty toward his 'superior' in terms of religious ethics, but because diligence and enterprise raised production, created an 'abundance of everything' and contributed to the pursuit of the fine arts and learning. The motives behind his reasoning were secular and temporal. Furthermore, Copernicus did not profess the idea of economic self-sufficiency as did Thomas Aquinas. On the contrary, he placed great stress on the importance of foreign trade, desiring to create favorable conditions for the expansion of that trade with good money.

All the ancient and medieval writers, as well as many authors of a later time, treated agriculture, that is ownership of land (not work on the land), as the most important and most dignified of all 'occupations'.* Copernicus made no single mention of agriculture. He did, however, point to the importance of 'first rate craftsmen' and of learning and the arts. The social ideal proposed by Copernicus was urban, based on the craft industries and production, with the productive efforts of the producers given first importance. Unlike the writers of the Reformation, he took a positive view of foreign trade although he most firmly opposed speculation and the tyranny of monopolies (as evidenced by the bread price regulations to which we shall return shortly).

New elements, born of the Renaissance, entered the Copernican concept of 'abundance' as well as his ideas on productive employment, on the struggle against indolence and on beggary. One of the fundamental components of mercantilism was the creation of the economic prosperity of a country (an abundance of goods) and employment of the available labor force (eradication of indolence). One of the first mercantilist treatises, *A Discourse of the Common Wealth of this Realm of England* (1549), attributed to John Hales, an Englishman, postulated that the state ought to take recourse

* Probably Erasmus of Rotterdam alone does not recognize the dominance and priority of agriculture over other trades (see: *Roscher Geschichte der National Oekonomic in Deutschland*, Munich 1874, p. 41). Erasmus also speaks with contempt about the merchant: 'the guild of merchants is the most foolish and the most foul because their occupation is also the foulest and carried out by the foulest methods'. *Praise of Folly*.

to any measures that would insure an abundance of goods and this called for hiring help in agriculture and for employment in towns of all the people that were fit for work.* Copernicus failed to place stress on the active role played by the state in the area of production and employment. Copernicus expected the state to play an active role only in the restoration of currency.

The age of mercantilism and of primary accumulation emerged after the Renaissance. The idea of an 'abundance of goods' to satisfy the needs of wealthy people was replaced by the dominating idea of economic expansion and by the postulate of a favorable balance of trade which brings an increase 'in wealth' in the form of gold and silver. Manufactures sprang up and the industries began to produce primarily for the mass market and not for the manors and the palaces. Copernicus did not belong to that age. He was a man of the Renaissance in every respect.

Of interest are the great astronomer's social views which have come down to us in documents connected with his activities as administrator of the Warmia chapter.

The bread price regulation (*panis coquendi ratio*), issued by him in 1531, established the price of bread in Olsztyn and in other localities of Warmia. Bakers were known to take advantage of the fluctuations in the price of grain to make an additional profit. The price regulation lists in detail the relations between the available supply and the price of flour and bread to the price of grain. 'From this,' the regulation states, 'one easily finds the true and just weight and price of bread.' The weight must be exact, the usual practice of merchants to underweigh must no be tolerated 'because we do not seek to profit but to give an accurate measure' (*non mercaturum sed certum modem inquirimus*). The price of bread, according to Copernicus, should include the cost of grain and the baker's salary. It should therefore be determined principally by labor costs and expenses and should not include the profits of the monopolist. The scholastic concept of 'a just price' (*justum pretium*), which was the original definition of value according to labor, was accepted by Copernicus. The concept of 'a just price' was taken from scholastic moral teachings which also represented the political economy of feudalism. Despite their religious context, the teachings reflect the relations that prevailed in the area of production at the time. Expanding production of goods and the law of value peculiar to it destroyed production relations that had been established over the centuries and that were sanctioned by tradition. At a time when feudal relations in production still remained in harmony with the early level of productivity, the desire took hold to keep in check the unbridled forces of the production of goods, which took the form of trade capital, by enforcing the observance of the moral and religious precepts of 'a just price'. The just price covered labor costs and expenses (*labor et expensae*). Thus, the effort to maintain an equivalence of exchange was made; it was derived from the principle that equality in labor costs was the basis of exchange. Adherence to the just price policy was seen as a means of curbing monopolistic ambitions and of restricting the exploitation of the rural area by the towns; it was conceived

* J. Hales: *A Discourse of the Common Weal of this Realm of England*, Cambridge 1893, pp. 59 and 98.

as a precautionary measure guarding against abuse on the part of the 'natural' monopoly held by the craftsmen and specifically by the shopkeepers. Owing to the poor condition of roads and transportation only a tenuous contact was maintained between the individual local markets. Consequently, the shopkeeper selling salt or cloth to the consumer enjoyed a kind of natural monopoly and could easily cheat and take advantage of the buyers. Full freedom in establishing prices would have destroyed the relative stability of the prevailing economic system.

At the time in which Copernicus lived conditions were still favorable for the natural monopoly enjoyed by the craftsmen and the shopkeepers with regard to the buyers. The growth of market trade, the fact that the individual had to buy increasing quantities of commodities at the market, further strengthened that dependence because productivity grew at a slow rate and the supply of goods was too small to undermine the domination of natural monopoly.

Under the circumstances, the promulgation of price regulations was both necessary and justified. Later, when the age of primary accumulation was at its peak and when economic concepts and principles of economic policies were formulated by representatives of commercial and money-lending capital, the principle of a just price was gradually replaced by a new principle which said:'that price is just which is determined by the market' and so price regulations of any kind are unnecessary and harmful. The new economic ideology, which was an expression of the age of primary accumulation, was to emerge and to predominate in the 17th century.

Copernicus broke radically with the medieval world of concepts not only in the science of money, he rejected out of hand the feudal 'dominal' concept, but he was also 'modern' (*bürgerlich*, as Marx said), lay and temporal in his ideals of society. Copernicus divorced himself from the method employed by scholasticism; he did not quote, voice or ruminate on the thoughts of the Church Fathers but instead relied exclusively on observation of facts, evolving economic laws from his observations.

The category of social production, the concept of 'prosperity' and the category of productive labor were unknown to medieval scholastic thought. Feudal relations in production based on privilege, land ownership and the duty to render services in kind on the part of the producer created in the religious superstructure the concept of work as duty and the concept of diligence as a moral command. They failed, however, to provide a foundation for the development of the concepts: productive labor, the sources of 'wealth' and the importance of the production process. The 'labor' of people, who were compelled to work because they owned no land, constituted not an economic but a moral problem expressed by the command and the duty to be diligent; the aim of these commands was to defend the established social order. The salvation of the soul was of the utmost importance. Toward that end it was necessary to pray and to work. Wealth and property were given in order to be used in doing good works for the benefit of one's fellow man, and in helping people deprived of resources and of the means of a livelihood. In order to insure to the privileged classes a secure enjoyment of their wealth, the ideal of poverty was proclaimed as indispensable to the salvation of the soul; it was the rich man's duty to 'help the poor'.

None of these categories are found in Copernicus. On the contrary, when he speaks of the decline and of the rise of 'kingdoms, duchies and republics', he gives priority to economic questions. He stresses the influence of the 'import of foreign goods and of trade', the importance of works of art, of first rate craftsmen and of 'the abundance of goods' to satisfy needs and describes the bad effects of sloth, indolence, neglect of the fine arts and learning and of the shortage of basic goods. Copernicus valued highly the merchants who traded in foreign goods and distrusted the greedy craftsmen and the small shopkeepers.

While there is no doubt whatever that Copernicus was first in formulating the modern theory of money, he cannot be considered first in the analysis of the economic bases of social evolution. The first and relatively systematic analysis of the processes of production is found in St. Anthony who lived in Florence in the 14th century. The city based its economic existence not on feudal ownership of land but on a highly developed capitalist system of industrial production, international commerce and banking. However, the salvation of the soul was most important to St. Antonin. He contemplated economic questions in the context of religion and its moral commandments.

One may risk saying that had Copernicus published his views on economics in a more elaborate form, on the order of *De revolutionibus*, he would have been acclaimed the originator of the economic theory of early capitalism, which by its precision of thought and contents was unsurpassed by any of the first economic analyses of the Renaissance.

The memorial written by Copernicus was and remained unknown in Europe until the 19th century. The paper was not issued in print in his time, except for a summary published in 1592 by Kasper Schütz in his *Historia rerum Prussicarum* (book X, pp. 517–520).

The ideal of a society which enjoyed an abundance of goods, where sloth and indolence were absent, which had first rate craftsmen and so on was an intellectual elaboration of those elements that were rooted in the relations observed directly by Copernicus. He did not reject the unfolding reality, he did not deny it and did not create a picture of a rational and sensible, visionary and unfeasible reality as did his contemporary, the great humanist Thomas More in his *Utopia*. Copernicus sought to discover the ways that led toward a society which was feasible, tangible and verifiable. He wished to build a society organized on rational principles by developing the features that already existed in it. Copernicus idealized the society that was in the making, while More described a plan for a society organized on rational principles by way of negation and criticism of existing reality. But Copernicus and More, as well as Erasmus of Rotterdam, had the same historical background and represented similar scientific attitudes and a lay cast of mind; they had also freed themselves of the influence of scholastic thought.

Every new social formation adopts some of the economic categories of the disappearing formation. It is the social contents of the categories that undergo a change. The principle of *iustum pretium* did not perform the same function in Thomas Aquinas as it did in Copernicus. The purpose of the principle of 'a just price' in scholastic theory was to strengthen the feudal base, the feudal system of production. The purpose of the *iustum pretium* in 16th century economic thought was to protect the masses of the population from the ruinous effects of the invasion of capitalism at a time when productivity was weak and the economic existence of large masses of people insecure and unsound and when the new production relations (monopolies, marked speculations) ruined large masses of the population economically. Monopolies were a manifestation of progress, for they concentrated the means of production. However, at this stage of the development of productivity, the monopolies in their effect represented a threat to the entire society. The protection of the rural area from exploitation by the town, the protection of the consumer from exploitation by the monopolist-producer and by the shopkeeper was as necessary as it was progressive in the same way as the growth of capitalism in the town was a progressive feature and as primary accumulation of capital was progressive. This is the dialectic of historical evolution.

Copernicus always referred to the state as a whole; he emphasized the interests of the entire republic when he proposed that two mints be designated in the state: one in 'the lands of His Royal Highness, the second in the lands administered by the Prussian prince', making one condition 'that both coinages be subject to royal authority' and that 'at the command of His Royal Highness they be accepted as currency in the entire Realm'. He added: "This will carry no mean weight in reconciling the minds and in easing trade relations".

These words reflect not only an understanding for the importance of the internal market but also speak of the patriotic allegiance of Copernicus to all of Poland and of his concern for the welfare of all the countries of the multi-national Polish state.

THE HELIOCENTRIC CONTROVERSY IN EUROPEAN CULTURE

BARBARA BIEŃKOWSKA
University of Warsaw

> Multos expertus est inimicos, quinimo etiam et post mortem

Every age has its problems. Every vigorous culture has its great controversies. Every new and valid theory of science goes through periods of indifference, indignation, incredulity, through phases of explicatory and militant propaganda only to be fully accepted in the end when, incorporated into the systems of conventional truth, it is so self-evident that it becomes virtually a cliche. Yet not many scientific theories of critical importance generated controversy or became the ruling issues of their age. One of the few that did was the heliocentric theory. The long and tempestuous controversy that raged around the heliocentric theory had as a crystal ball brought to a focus the fundamental contradictions immanent in European intellectual culture of the late 16th and of the early 17th century. However, not only the fundamental scientific and ideological concepts but also the most commonplace ideas had to be revaluated before the heliocentric theory could win acceptance. Adherence to geocentrism soon lay beyond the pale of the intellectual mainstream. Thus the choice faced at first by astronomers was soon confronted by whole generations of the artists and recipients of European culture.

The heliocentric theory represented a direct challenge to the three pillars of knowledge about the world: the Bible, Aristotelian philosophy and sense perception.

Most difficult to reconcile were the contradictions between the basic principles of the heliocentric theory, i.e. the motion of the Earth, and certain passages of the Holy Scriptures.

The most frequently cited passages appear in: the Book of Joshua 10:12–14 stating that at Joshua's request God commanded the Sun to stand still so that Israel could avenge themselves on Gibeon; in Ecclesiastes 1:4–6 where Solomon wrote,

> the Earth abideth for ever. The Sun also riseth, and the Sun goeth down and hasteth to his place where he arose. The wind goeth toward the south and turneth about unto the north;

in Psalm 93 verse 1 –

> the world also is stablished that it cannot be moved,

in Psalm 104 verse 5 –

> Who laid the foundation of the Earth, that it should not be removed forever;

finally in the Second Book of Kings 20:11 –

> And Isaiah the prophet cried unto the Lord: and he brought the shadow ten degrees backward by which it had gone down in the dial of Ahaz.

The contradictions were indeed indisputable when the above (and similar) passages of the Bible were taken literally. Furthermore, in view of the all-encompassing authority that the Scriptures exercised on the consciousness of the population at the time, any discussion or thought about the motion of the Earth was totally inadmissable. It was principally in this context that, earlier than the representatives of the Catholic Church, the leaders of the Reformation (Luther and Melanchthon) condemned the heliocentric theory as blasphemous.

In the face of the incontestable scientific value of the new theory, precluding any a priori rejection, scientists, most of whom were devout Christians, sought to reconcile the two truths: the scientific truth and the revealed truth. The way out of the dilemma could have been (and indeed became in time) the allegorical interpretation of the Scriptures. Advocates of this approach called to witness St. Augustine when they asserted that since the authors of the Bible addressed themselves to the common people they had to resort to simple and colloquial formulations that appealed to the imagination and were readily understood by them. Speaking convincingly of ordinary matters they wished to arouse the confidence of the people in the important question of faith and morals. The Bible speaks of the Heavens as they seem and not as they are in fact. 'The Holy Scriptures tell us how to go to heaven and not how to measure the heavens. The Holy Scriptures wish to make Christians and not astronomers'. These succinct arguments were frequently advanced by the heliocentrists of the 16th, 17th and even of the 18th century.

The allegorical and metaphorical interpretation of the Scriptures followed two basic lines of reasoning. First of all an effort was made to interpret the known Biblical passages in a variety of often complicated and ingenious ways with the purpose of proving that, in their deeper and metaphorical sense, they actually speak in support of the heliocentric theory or at worst do not contradict it. Other passages, which in their literal sense contradict the Ptolemaic theory (geocentrism), were eagerly quoted. The second and more independent approach was to state that the Bible could not be considered as a reliable source as far as scientific questions were concerned, that these were treated superficially and were presented in a manner readily accessible and consequently acceptable to the simple people and finally that none of the formulations found in the Bible can be taken as valid arguments for or against the heliocentric theory because they are not scientific assertions.

Indispensable to the development of science the concept of an allegorical interpretation of the Holy Scriptures, reverting to St. Augustine, was evolved in the 14th century by Nicole Oresme. His interpretation closely resembled the views of Copernicus himself. Copernicus stated explicitly in the dedication of his work to Paul III that he feared attacks upon the heliocentric theory on the part of those who would quote certain passages from the Bible 'falsely distorted for their purpose' *male ad suum propositum detortum*). Although Copernicus realized that his theory contradicted several fragments of the Bible in their literal context, he was convinced at the same time that the allegorical interpretation was the valid one. This line of reasoning was later elaborated by Wright, Galileo, Foscarini and by the overwhelming majority of the educated champions of heliocentrism. This most hotly debated point would not have been contested had the Christian Churches accepted in the early 17th century the allegorical interpretation of the cosmological passages of the Holy Scriptures, an idea that was popular with the intellectual circles at the time. But from 1616 on, the Catholic Church rallied all the weight of its authority to the defense of the literal interpretation of the Scriptures. It was not until 1757 that under the pressure of incontrovertible scientific proof Pope Benedict XIV finally issued a decree which softened the position of the Church on that issue. But the idea of an allegorical interpretation of 'the scientific' passages of the Bible was not recognized as the official doctrine of the Catholic Church until 1897, during the reign of Pope Leo XIII.

As a consequence, for more than two hundred years the Catholics found the road to a reconciliation between the discoveries of science and the official teachings of the Church totally cut off, or at least fraught with obstacles. The decrees of 1616 and 1633 banned not only the work of Copernicus but also all the subsequent works based on the theory of heliocentrism, although it was difficult to find even then (and much more difficult in later years) a serious work of astronomy based on the geocentric principles.

Representatives of the Protestant Churches continued their opposition to the allegorical interpretation of the Holy Scriptures throughout the 16th and in the early 17th century. Protestant opposition, based on Biblical grounds, was as violent as that of the Catholics, although it was less official and authoritative in character and for that reason was more easily vanquished by the open-minded theologians of the various heretic creeds.

Owing to its innate conservatism, the Orthodox Russian Church came out last against the heliocentric theory (the problem did not arise within its sphere of influence earlier). Its opposition, made manifest as late as the first half of the 18th century, was also the most virulent. Yet its belated action was of no greater significance.

The only way out of the conflict was a compromise solution whereby the heliocentric theory would be, honestly or to all outward appearances, recognized as a hypothesis. It could then be popularized and developed without leading to open conflict with theologians and the defenders of common sense.

The version, holding that the heliocentric theory, as all other cosmological systems of the past, is a hypothesis was advanced by Osiander in the preface attached to the first 1543 edition of *De revolutionibus* and included again in the editions brought out in 1566 and 1617. The preface was printed without Osiander's name appended to it so as to imply that it had come from Copernicus himself.

Osiander's preface describes Copernicus' theory as yet another hypothesis – as an arbitrary mathematical construction which is all the more useful and worthy of admiration because it is understandable and because it is related to a rich store of relevant observations. Osiander therefore applied to the work of Copernicus a stereotype which was repeated in every manual of astronomy published at that time. Hypothesis was defined as a theoretical premise by which astronomers sought to explain the celestial phenomena and their own observations in a way that would enable them to formulate theories which they then could present to their pupils. The more perfect the hypothesis, the simpler the explanation of the motion of celestial bodies and the easier the calculation (calculus). However, one should not seek to discover in a hypothesis any reflection of reality or any probability.

Contrary to prevailing assumptions, to which Osiander also gave expression, the idea that the heliocentric theory is one of many hypotheses about the structure of the universe, or that the science of astronomy is incapable of discovering or of formulating a theory on the truth about the universe, was entirely alien to Copernicus.

Copernicus made that perfectly clear in the dedication of *De revolutionibus* to Pope Paul III. He was still more explicit on the point in the original preface which, omitted in the early editions, was finally issued in print in the Warsaw edition of 1854. Copernicus wrote there, among others:

But this divine rather than human science, which inquires into the highest matters, is not without difficulty, particularly as we find that those who entered upon its pursuit did not for the most part agree among themselves regarding the basic assumption, called hypothesis by the Greeks, and for that reason did not base themselves on the same principles (Porro divina haec magis quam humana scientia, quae de rebus altissimis inquirit, non caret difficultatibus, praesertim quod circa eius principia et assumptibus, quas Graeci hypotheses vocant, plerosque discordes fuisse videamus, qui ea tractature aggressi sunt, ac proinde non eisdem rationibus innixos).

That the term hypothesis came to mean a tentative and unproved proposition in the literature of later years was due to the consensus held on science and on the role of astronomy at the period discussed here.

Yet the hypothetical acceptance of the heliocentric theory, contrary though it was to Copernicus' convictions, was scientifically justified until about the mid-17th century, that is until its validity was proved in the physical sense. Continued recognition of the heliocentric theory as a mere hypothesis, upheld strenuously by many of its enemies and buttressed by the highest authorities on the sole ground that it did not accord with the Bible, was purely ideological in nature from the 1690's when scientific arguments began to speak convincingly in support of the validity of the new system.

The damaging label 'hypothesis', attached to the Copernican theory, nevertheless opened a way for the wide dissemination of the main principles of heliocentrism. This fact exerted a marked effect upon the educational processes after 1616 and was to prepare the ground for the acceptance of the new cosmology by the educated public.

The traditional – Aristotelian-Thomist – cosmological vision which took form in the Middle Ages was an integral part of knowledge about the world and about man. The system accepted such general principles from theology as: God created the Heaven and the Earth; He imparted motion to the Heaven and made the Earth stand still. Philosophy, on its part, was encumbered with the task of explaining the mechanism of the motion of the heaven, the structure of the heaven and the place occupied by the Earth in this system. In a sense the celestial system was divine because the seat of God was located at its highest point. The mechanism of the heaven was governed by the laws of Aristotelian physics. The Earth, heavy and immovable, was located in the center of the universe. Closest to it was the heaven (*caelum*) of the Moon, followed by the 'heavens' of Mercury, Venus, the Sun, Mars, Jupiter and Saturn, by the heaven of the fixed stars, by the crystalline heaven, the '*primum mobile*' and finally by the

Lord's Heaven, the seat of God, of the saints and of the saved. The 'heavens' were the spheres along which, it was believed, the individual planets revolved in a uniform motion. The '*primum mobile*' revolved at the greatest speed and it was this sphere which allegedly influenced the motion of the spheres within it. The spheres were real and transparent but impermeable crystalline barriers. The ideal space of heaven was unchanging and immaculate in contrast to the base and transitory earthly space. This concept of cosmological space invariably gave rise to obvious consequences in world outlook.

The powerful influence exercised by this suggestive vision of the world was further enhanced by the fact that it accorded with the sense perceptions of every individual. 'Our eyes bear witness to the revolution of the heavens', Martin Luther argued with indignation. The Copernican picture of the Earth, not perceptible to its inhabitants, seemed utterly preposterous from the practical point of view. Everyone was fully sensible of the stability of the Earth and of its central position. Everyone could see that it was the heavenly bodies which rotate around the Earth. Meanwhile the Copernican theory discarded the whole well-ordered body of knowledge and denied the immutable, as it seemed, laws of physics. It contradicted man's experience and common sense and, most important of all, challenged the validity of the formulations of the Bible. That was the substance of the pronouncements contained in the decision issued by the censors of the Holy Office on February 24, 1616. It was stated therein that the view that the Sun rests immovable in the center 'is mad and absurd'. The decision was followed by a decree issued by the Congregation of the Index of Prohibited Books (*Index Librorum Prohibitorum*) on March 3, 1616, which forbade the Catholics to read the work of Copernicus until it was corrected to define the heliocentric theory as a hypothesis.

The proposition put forth by Copernicus in 1543 in *De revolutionibus* did not resolve but raised new mathematical, physical, methodological and above all ideological problems. That is why it gave new direction to the scientific research of his period, that is why it sparked the understandable and long lasting controversy.

1. The Early Phase of the Heliocentric Controversy (1543–1616)

The work of Copernicus soon attracted attention. It aroused the interest not only of astronomers but also of a large group of historians, antiquarians, lexicographers, bibliographers and biographers. Their opinions and assertions served to spread the reputation and the greatness of Copernicus throughout many countries and at the same time provided proof of the wide knowledge of the life and work of the Polish astronomer. A few years after *De revolutionibus* appeared in print, Copernicus was widely recognized as one of the greatest scientists of all time. That opinion remained unaltered regardless what the attitude toward the heliocentric theory. Mention of Copernicus and of his work was made not only in scientific literature but also in the then widely disseminated and widely read chronologies of world events, 'portraits' of famous persons, anthologies of proverbs and bibliographies, to take the example of the famous *Bibliotheca universalis* of Conrad Gesner, brought out in Zürich in 1545. Copernicus was described there as the reformer of astronomy, as an unrivaled mathematician and as a consummate Sarmatian (as the Poles were sometimes called) scholar. Certain informative publications also gave short and objective descriptions of the principles of the heliocentric theory (as for example the *Opus chronographicum*, by P. Opmeer, published in Antwerp in 1617). In this early period, however, the heliocentric theory did not yet play a larger role in the general consciousness of the educated population. Why even in the eyes of the overhelming majority of astronomers the detailed observations and calculations, which immediately won their respect, were more important than the risky theory of the solar system, often treated as an unimportant idiosyncracy. Hence, the astronomer's mathematical results were accepted without any mention of the principles on which they were founded. For example, E. Reinhold in compiling his *Prussian Tables* in 1551, a most remarkable popular work of its time based on *De revolutionibus*, did not even mention the heliocentric system.

That is how matters stood in practical astronomy for a fairly long while. It is true that a small number of astronomers, who conducted their research with imagination and who, embroiled in the contradictions inherent in Ptolemaic geocentrism which the compromise system of Tycho Brahe (1546–1601) was incapable of resolving, took the heliocentric concept as the point of departure for their studies, did come up with remarkable results. Consequently, they also undertook to explain the basic principles and to disseminate the findings of the theory. The widely read brochure on the subject, published in 1576 by the British mathematician Thomas Digges, *Perfit description of the caelestial orbes*, ran into six editions by 1605. The brochure gave an explanation and an apologia of the Copernican theory, treated as a true picture of nature.

The following exercised a decisive influence on the course of the scientific discussion: Kepler's well known activity (1571–1630), his astronomy was based on the hypotheses put forth by Copernicus (a fact which he makes clear in his works) and Galileo (1564–1642) in his research and role as popularizer, whose achievements are closely related to the Copernican theory. The works of Kepler and Galileo tipped the scales of the scientific argument. From then on astronomy and its development were based on the heliocentric theory, conceived not as an alternative but as the only course to follow. The compromise systems (such as those devised by Riccioli and Longomontanus) were created for teaching purposes and not for the use of science. Dictated by reasons of ideology, they represented an effort at a reconciliation between the immobility and the central position of the Earth and the detailed findings of the new astronomy.

Consequently up to the mid-18th century and even longer these compromise systems were taught in schools which, constrained by religious injunctions, could not propagate the heliocentric theory of Copernicus.

Fascinated by the striking disclosures, by the improbability and by the philosophical consequences of the new hypothesis, a small number of poets and thinkers, like the small group of astronomers, evinced an interest in heliocentrism. However, to the majority of these individuals, marked by great impressionability and imagination and by a minimum of mathematical knowledge, the Copernican concept appeared terrifying in its absurdity. They attacked it principally with arguments of common sense which carried conviction with those who had no knowledge of astronomy. Despite the strenuous efforts of scientists to explain that the evidence of the senses is not, nor can it be, the ultimate criterion where the corroboration of scientific theory is concerned, these arguments were repeated in the schools for a good many years.

It is from this position, taking his arguments from daily experience that, as early as 1578, the French poet Du Bartas fired a blast against the heliocentric theory in his *Week of the Creation of the World.* Soon afterwards the Scottish poet George Buchanan took up the polemic with Copernicus in *De Sphaera libri* V, published in 1591, arguing that if the Earth moved then all the objects found on it would topple. The English poets John Davies (1594) and Joshua Sylvester (1604) were attracted by other aspects of the new theory. They described and praised the heliocentric system in their poems. Remarking on the growing popularity of the new cosmological system, John Donne (1611) expressed a concern for the philosophical consequences and effect on world outlook that the theory would have. The famous English poet John Milton, writing more than fifty years later (1666), displayed less foresight. Describing the two opposing planetary systems in '*Paradise Lost*', he did not commit himself as to which he believed to be the valid one but treated the problem as an abstract scientific controversy.

More comprehensive, though equally dilettante in nature, was the analysis of the heliocentric theory given by Jean Bodin, famous French historian and political writer, in his *Universae naturae theatrum*, published in 1592. According to Bodin, Copernicus was wrong because, if it is accepted that the Earth has three motions and if we add to them the gravitational motion toward the center, then it must be admitted that the Earth has four natural motions, and these assumptions are contradicted by all the physical principles which attribute only one motion to everybody. If the Earth, heavy and inert, turned around its center and around the Sun then the houses, cities and mountains would crumble with every motion. The theory also is contrary to the Bible which says that the Earth abides for ever. Bodin fought the heliocentric theory with arguments corroborated by daily practice and by Aristotelian physics. He mentioned the inconsistencies with the Bible only in passing and without giving them special emphasis. Soon, however, it was these objections that gained primacy owing largely to the activity of Giordano Bruno.

This philosopher and mystic (burned at the stake in 1600 for his theological views) took the heliocentric theory as the framework of reference for his fantastic cosmological concepts. Working from the heliocentric theory he wove, with the evocative power and talent a of poet and a visionary, prospects of an infinite and immeasurable world, prospects of great worlds, which were to fascinate the 17th and 18th centuries and to increase the fame of the Copernican theory. Simultaneously, he drew attention to the ideological dangers inherent in it which mobilized theological opposition. From the theological point of view, the idea of the greatness of the worlds was fraught with greater dangers than the theory of the motion of the Earth because it not only contradicted many passages from the Old Testament but also contravened the whole idea behind the New Testament.

It is not surprising that the champions of the heliocentric theory began to focus their arguments on this point. Most significant in this respect was the preface written by the English astronomer E. Wright to the famous book on earth magnetism by William Gilbert (*De magnete*, London 1600). In Wright's opinion the diurnal motion of the Earth provided a perfect explanation for all the astronomical phenomena which could not be explained earlier. Contrary to the asseverations of the opponents, logical reason also spoke in favor of it: it is easier to assume that the small Earth rotates around its own axis than to imagine that the whole vast universe revolves at a terrific speed around the Earth (*Frustra fit per plura, quod fieri potest per pauciora*, Wright wrote). Nature does not commit such mistakes, for she is economical and thrifty. This argument (first outlined in Copernicus' dedication to Paul III) was to be invoked by the advocates of heliocentrism over many decades. It was sometimes given in the form of an anecdote about a cook who, turning a roast on a spit in the fire and not the other way around, resolved the controversy of the astronomers.

The theme presented by Copernicus, that is the frugality of nature, the purposefulness and the harmony of the structure of the Universe, were to be raised with growing frequency by supporters of heliocentrism as an ideological argument. Copernicus, Wright, Bruno, Galileo, Kepler, Bérulle, Muratori, Śniadecki and a whole galaxy of other heliocentrists of various creeds were to steadfastly elaborate the idea that the heliocentric theory has finally revealed the extent of the power and of the wisdom of the Creator by disclosing the perfection of his works. The idea was finally embraced by the educated circles about the mid-18th century, reconciling the by then unshakable achievements of modern science with the authority of theology which by then tacitly accepted the allegorical interpretation of the Bible as an ineluctable fact. However, before the structure of the new physics, the new cosmology and the new religious interpretation was raised about the heliocentric theory, the mounting contradictions led to a crisis. *De revolutionibus* was placed on the Index of Prohibited Books in 1616. With this fact the heliocentric controversy entered upon a new phase in Europe.

2. The High Period of the Scientific and Ideological Dispute on Heliocentrism (1616–1687)

Scholars, who were gradually accumulating evidence in support of the Copernican system, did not succeed to put an end to the dispute with incontrovertibly convincing arguments although the number of arguments continued to grow. It was finally Isaac Newton, who turned the scales in favor of heliocentrism when his *Philosophiae Naturalis Principia Mathematica* appeared in 1687. From that day on Newtonian Mechanics was to be invariably associated with Copernican astronomy. That is why the year 1687 has been taken as the closing date of the high period in the scientific controversy. True, many decades were to elapse before the discoveries of Newton and of his immediate predecessors were assimilated by all scientific and educational centers and before the controversy over the essential merits of heliocentrism subsided. Nonetheless, by the 1690's the heliocentric theory became a fully substantiated scientific thesis and to speak of it as of a hypothesis after that date was taken as proof of scientific conservatism or of the complete subordination of scientific matters to the decision of the Congregation of Cardinals.

In the late 17th century the dispute had clearly shifted to ideological grounds and it began to implicate far larger numbers of recipients of scientific discoveries and cultural products than of scientists and of artists working in the field of culture.

The emergence of these new circumstances bears a direct relation to the famous case of Galileo.

In 1610 Galileo published the dissertation *Nuntius sidereus* where he described what he had observed with his telescope. His findings provided new arguments supporting heliocentrism. One of the most important arguments was his discovery of four satellites or moons of Jupiter and his observations of the phases of Venus which suggested that the planet revolved about the Sun. In 1613 Galileo observed with his telescope the shifting of the spots on the solar disc. Apart from the fact that his discovery of spots on the Sun delivered a blow to the thesis which held that the celestial bodies were perfect and unchanging in their nature, the shift of the spots indicated that the Sun rotated around its axis, which was equivalent to the diurnal motion of the Earth. Galileo did not keep his discoveries secret. On the contrary, he made every effort by way of publication, conversation, meetings and correspondence to win the largest number of adherents to these truths. He assailed his opponents with a vehemence and passion that left them baffled and often infuriated them. Galileo's opponents were professors – Aristotelians – who refused to be reconciled with the new physics, and the Catholic clergy (chiefly Dominican friars) who accused Galileo of voicing theories which were inconsistant with the Holy Scriptures. Opposition against Galileo grew so strong in circles close to the pope and the College of Cardinals that he decided to make an explication of his position. He did so in a letter written to Benedetto Castelli in 1613 and further elaborated his arguments in 1615 in what has come to be known as the letter to Christina, duchess of Tuscany. In both letters Galileo argued that the Bible's references to the system of the Universe were allegorical (metaphorical). When the references are interpreted in this manner, we then find that the actual state and order of nature – the handiwork of God – is not inconstistent with the Holy Scriptures, which were also dictated by God, for great truths cannot exclude each other. God is the author of two books – the book of nature and the book of the Scriptures. God cannot contradict Himself.

A similar position was taken by Foscarini, the Neapolitan Provincial of the Carmelite Order, in a brochure published in 1615, which was a tract on the interpretation of Biblical passages that pertained to the heavens and to the Earth. Both Galileo and Foscarini professed their loyalty to the Catholic Church. Both treated the Holy Scriptures as the word of God and at the same time both were convinced of the validity of the heliocentric theory.

Foscarini sent his tract for evaluation to the Jesuit, Cardinal Roberto Bellarmino, renowned in Rome as an opponent of innovations in theology as well as in science. Bellarmino replied to Foscarini in a letter dated April 15, 1615, representing by the same the position of the highest Church authorities. The cardinal stated that he had always understood Copernicus' theory to be a hypothesis which provided a better explanation of celestial phenomena than the Ptolemaic system. In this context the heliocentric theory did nor represent a danger to faith and was serviceable to mathematicians. Acceptance of the theory as a truth of nature aimed against scholastic philosophy and theology. Even if proof was available (and Bellarmino believed it was not) that the Sun was *in fact* in the center and that the Earth did *in fact* revolve about it, it would still be necessary to exercise great caution in interpreting the Biblical passages which stated otherwise. When in doubt, the literal interpretation recommended by the Fathers of the Church should not be discarded. Nor can the illusion that the shores seem to recede when seen from a sailing ship be transferred to all the celestial phenomena. The evidence of the eyes clearly demonstrates that the Earth stands still and that the Sun, the Moon and the stars revolve.

In his reply to Foscarini, and indirectly also to Galileo, Bellarmino defended not only the principle of the literal interpretation of the Scriptures, he also defended traditional physics and sense perception as the only confirmation of the validity of scientific theories.

Consequent upon Bellarmino's assertions came the decision of the Commision of Censors of the Holy Office, issued on February 24, 1616, and the decree issued by the Congregation of the Index of Prohibited Books on March 3, 1616 forbidding Catholics to read the work of Copernicus until the correction defining the heliocentric theory as merely a hypothesis is introduced.

Informed about the decision of the Church authorities, Galileo is said to have promised solemnly to refrain disseminating under any form the theory of the motion of the Earth.

Cardinal Maffeo Barberini, known for his friendly interest in the new learning, became Pope in 1623. Galileo decided to speak out once again in defense of the heliocentric theory. Before he did so, he first spoke with the new pontiff. Urban VIII advised him, among others, to insert a preface and a conclusion in the book he planned explaining therein that the heliocentric theory was merely a hypothesis. To all intents and purposes Galileo seemed to have complied with these demands. *Dialogues on the Two Principal Systems of the Universe*, published in 1632, did contain a preface and a conclusion where the author alleged that the heliocentric theory was presented as possible. The body of the book, however, provided proof that wholly supported the validity of the Copernican theory. The Inquisiton was not hoodwinked that easily. The events that followed are well known: Galileo, then in his seventies, was brought to trial for violating the injunction of 1616. On June 22, 1633 Galileo was ordered to publicly recant his views and was sentenced to perpetual surveillance. His book was prohibited and the heliocentric theory was once again condemned.

The injunctions contained in the decree of 1616, at whose roots lay the formal inconsistency between the heliocentric theory and the Bible as well as fierce personal antagonisms, were redoubled in force in 1633. They were to weigh heavily with the Catholic world as regards its position toward Copernicus and his successors.

The ban issued by the Congregation of Cardinals could not stop the astronomers for long from creatively developing the heliocentric concept nor could it halt the publication of a growing number of books which furnished proof of its scientific validity and of the untenability of the Biblical rebuttals. The publications evoked a veritable avalanche of replies from the defenders of the decisions taken by the Congregation and, this ought to be made clear here, the defenders were not all members of the Catholic Church. In this period marked by fierce ideological battles, the Protestants often made capital of the ban imposed by the Holy Office, using it as evidence of the restraining influence the Catholic Church exercised on the development of science. In their assaults on heliocentrism Catholic writers on their part were only too eager to present it as a theory advocated by heretics. On both sides, however, these arguments were no more than unfounded propaganda gambits.

Although the heliocentric theory was in fact propagated quite freely among the Protestant communities, yet some of the leading defenders of that system were Roman Catholics (notably Descartes and Fontenelle). They, however, were often compelled to mask their views by all manner of subterfuges, quite easily deciphered, it might be added, by their contemporaries. The dividing line ran between the mathematicians and the theologians of various creeds. It is no wonder then that polemics between astronomers and theologians became the rule.

And so the theologians L. Fromondi and M. Inchofer came out against the pro-heliocentric publications by the astronomers Philip and Jacob Lansberg, A. Deusing, and D. Lipstorp, offering the well known and oft repeated arguments that the theory was inconsistent with the literal interpretation of the Holy Scriptures (no other interpretation was permissible) and with Aristotelian physics and that it was contrary to the evidence of the senses.

There were intellectuals, these principally Jesuits (i.e. A. Kircher, K. Scheiner, J. B. Riccioli) who, motivated by ideological considerations, sought to relate the details of the new observational astronomy with the general geocentric concept. But growing numbers of free-minded astronomers came out in favor of the heliocentric system arguing that the advances made in science during the last century not only failed to disqualify the Copernican theory but on the contrary fully supported it with a growing number of ever more convincing proof.

But ideological pressure was so great at the time that even the strongest personalities in science were compelled to resort to camouflage. A typical example is the attitude taken by Descartes in his *Principia philosophiae*, published in Amsterdam in 1644.

The great philosopher only partially recognized the validity of the heliocentric theory. He conceded that the Sun was in the center in relation to other planets but denied motion to the Earth. He maintained that the Earth, like other planets, is carried by the fluid matter of the heavens in what are the famous Cartesian vortices. Clearly

Galileo's trial was decidedly responsible for the hypothetical formulation of heliocentrism and for the apparent opposition to Copernican views on the motion of the Earth.

Descartes, writing to Mersenne in the well publicized letter of November 1633, stated that the heliocentric theory was the underlying foundation of his philosophy and asked Mersenne to inform him what stand the Inquisition takes on the subject. When Descartes learned of the condemnation of Galileo, in the very next letter to Mersenne, dated April 1634, he begged him to withdraw from print a work of his called *Le monde* where he openly propagated the theory of the motion of the Earth. He decided to evade censure by hiding his true beliefs.

Descartes' strategy was correctly interpreted by the two warring sides. Both, in total accord this once, recognized the philosopher, and accurately so, as a pillar of the heliocentric theory. Indeed, for it was principally on the ground of Cartesian opposition to Aristotelianism that the Copernican cosmology could be accepted and developed.

For the scientists of the second half of the 17th centrury, heliocentrism became an inseparable component of the scientific world outlook. For example Thomas Hobbes in his famous *Elementa philosophiae* (1655) accepted the Copernican planetary system quite naturally and without argument. But the controversy was not yet totally resolved. Robert Hooke, remarkable English mathematician, physicist and biologist, believed the dispute to be in full swing when in 1670 he wrote his treatise which appeared under the significant title *An Attempt to Prove the Motion of the Earth by Observations* (London 1674). Hooke quoted only scientific arguments, ridiculing disputants who, owing to their ignorance and prejudice, were incapable of even imagining the motion of the Earth. The majority of the educated men was aware of the weight religious arguments carried and so conducted the discussion on a broader plane. That was the course followed by Peter Megerlin, astronomer and mathematician of Basle, in his extensive volume called *Systema mundi Copernicanum argumentis invictis demonstratum et conciliatum theologiae* (Amsterdam 1682). He proved not only the physical validity of the heliocentric system but also its agreement, in the allegorical sense of course, with the Bible.

Yet the learned dissertations did not contribute as much to the popularity of the heliocentric theory as did the light and informal *Conversations on the Plurality of Worlds* (*Entretiens sur la Pluralité des Mondes*, Paris 1686), written by Bernard de Fontenelle, brilliant writer, author of works in popular science. Fontenelle took the Copernican theory as a point of departure for his reflections upon the hypothetical existence of other planetary systems, describing it as clear and logical and as concordant with the purposeful and economical structure of nature. Fontanelle repeated and elaborated the arguments put forth by Copernicus, Wright and others to the effect that the logical heliocentric system bore the most convincing witness to the wisdom of the Creator. This theme was to reappear more frequently later when it constituted an ideological interpretation of the by then accepted Copernican system. For more than a hundred years to come, Fontenelle's arguments lost none of their relevance. German translations of the *Conversations* appeared in 1789. By then 12 French editions, 4 English, 2 Russian and one Polish edition had been issued. The number of editions and translations provides the most telling proof not only of the brilliance of the Conversations, but also of the abiding attraction of heliocentrism (three generations of readers displayed an equally passionate interest in the problem). Furthermore, it testifies to how deeply engrained the traditional cosmological ideas were.

3. The Decline of the Heliocentric Controversy (1687 to the End of the 18th Century)

The achievements of astronomy and physics in Europe of the mid-17th century were crowned and codified by the work of Newton, which may be taken as the turning point in the scientific acceptance of the heliocentric theory. Geocentrism now vanished from science as an alternative.

The new vision of the world created by 17th century scientists did not remain confined to them. The general educated public, who carried out of school a picture of the universe as described by Aristotelian and Thomist philosophy, could alter that picture by reading popular science literature on the latest discoveries in the natural sciences and on Cartesian philosophy, which was so fashionable at the end of the 17th and in the early 18th century. Though they were free to change their views, the fact is that they did not always, or rather that they rarely, did so.

The heliocentric controversy continued unabated throughout the 17th and in the early 18th century, although it was now conducted on a different plane and level and involved different personalities in the dispute. To the 'independent' scholars and scientists, who were not connected with any school and who therefore were under no obligation to teach the outdated cosmological theories, the heliocentric controversy was pointless. The heliocentric theory was still defended by those to whom the reconciliation of the theory with the Bible was the most important objective.

Schools, particularly those operated by the monastic orders, began to give more attention to the heliocentric theory. Although its principles were explained, it was still treated as a hypothesis serviceable in astronomical calculations and its disagreement with a few passages of the Bible was the only reason given for not recognizing it as a truth of nature. In the last stage of the controversy efforts were made to demonstrate that the theory was in agreement with the Bible and to provide proof that it did not lead to atheism but, on the contrary, that by its perfection and harmony, it contributed to a better understanding and admiration of God.

Representative of these tendencies and important in its effect was the stand taken by the Catholic clergyman Lodovico Antonio Muratori, well known historian and religious writer of his time. Muratori belonged to the group of Italian priests who looked favorably upon the ideas of the Enlightenment. In *De ingeniorum moderatione in religionis negotio*, published in1714, the author oppugned the ostentation that crept into Catholic ritual and assailed the overstated interpretations placed upon modern learning. Appearing nearly a hundred years after Galileo and Foscarini, who voiced similar opinions, Muratori's work should be recognized as an important turning point in the controversy. In the course of the intervening years, demands that the 'scientific' passages of the Scriptures be interpreted allegorically were voiced solely by the heretics. Now for the first time in nearly a hundred years, a Catholic priest had spoken for the metaphorical exegesis of the Biblical passages and demanded that for the good of the Church the stigma of heresy be removed from the heliocentric theory.

Quite obviously, Muratori's opinion could not cause an immediate and a general turnabout in views or in the interpretation of the Bible. Many Catholics and many Protestants still continued to look upon the pertinent passages of the Bible as self-evident proof of the fallacy of the Copernican theory.

In the 1730's and later still, the heliocentric controversy continued to inspire new publications. But as the years rolled by these grew more and more anachronistic. The opinions they expressed clearly united heliocentrism with a whole complex of questions connected with Cartesian philosophy and modern natural science. The Copernican theory had by then become an inseperable element of the new scientific world outlook propagated by the Enlightenment. The resistance of conservative circles was invariably linked with opposition to modern learning and this was nearly unacceptable in the second half of the 18th century, owing to the level of the average scientific consciousness of the educated portion of society.

In this period too, the idea of an allegorical interpretation of the Bible suggested by Muratori began to win the support of the overwhelming majority of educated Catholics. It enabled them to overlook the contradictions between heliocentrism and the letter of the Scriptures. The decree issued by Pope Benedict XIV was a concession to this tendency. Catholic communities quickly accepted the new ideas concerning the universe and propagated each new achievement of astronomy and physics as proof of the wisdom of God and of the harmonious structure of Nature.

When this happened the great controversy on the Copernican theory was terminated.

SELECTED COPERNICAN BIBLIOGRAPHY

JERZY DOBRZYCKI

A relatively comprehensive catalogue of writings is to be found in the *Bibliografia Kopernikowska 1509–1955* (*Copernican Bibliography, 1509–1955*) by Henryk Baranowski, published in Warsaw in 1958, containing descriptions of close to 4000 titles of books and articles published up to 1955. An important supplement listing the literature published more recently is contained in the 'Adnotated Copernican Bibliography' in E. Rosen's *Three Copernican Treatises*, New York 1959.

A supplement to Baranowski's Bibliography, published in 1972, embraces the literature brought out since 1955. In the selection given below the numbers in parentheses correspond to the numbers listed by Baranowski in the 1955 edition.

1. Important Editions and Translations of Copernicus' Works

A. DE REVOLUTIONIBUS

The manuscript is found at the Jagiellonian Library in Cracow. Call number 10000.

Reproduction of the manuscript: (1) Gesamtausgabe... hrsh. v. F. Kubach, München. Ed. 1, *Opus de revolutionibus coelestibus manu propria*, Faksimile-Wiedergabe, 1944 (130) (2) *Opera omnia* Vol. 1, Facsimile, Warsaw 1972.

Latin Editions

(1) *De revolutionibus orbium coelestium libri VI*, Norimbergae 1543, J. Petreius (196); Reproductions: Paris 1927 (101); Amsterdam-Torino 1943 (126); Leipzig (copy from Kepler's library); Bruxelles 1966.

(2) *De revolutionibus orbium coelestium, Libri VI*, Basileae 1566, officina Henricpetrina (4); Reproduction: Praha 1971 (made from a copy owned by Tycho Brahe).

(3) *Astronomia instaurata libri sex comprehensa, qui de Revolutionibus orbium coelestium inscribuntur* (ed. by N. Mulerius) Amstelodami 1617, W. Jansonius.

(4) *De revolutionibus orbium coelestium libri sex* (ed. by J. Baranowski), Warsaw 1854 (27). (The title and text also in Polish. The preface to Book I by Copernicus, omitted in the earlier editions, was published here for the first time. The Warsaw edition also contains minor Copernican works.)

(5) *De revolutionibus orbium coelestium libri VI*, Curavit Societas Copernicana Thorunensis. Thoruni 1873 (45).

(6) Gesamtausgabe. München. Ed. 2. *De revolutionibus orbium coelestium libri sex* (ed. by Fr. Zeller and C. Zeller), 1949 (136).

(7) *De revolutionibus orbium coelestium liber primus* (ed. by R. Gansiniec and A. Birkenmajer), Warsaw 1953 (142). (Edited and translated into Polish: Book I to chapter 11 inclusively.)

Translations

Polish Translations

Translated by J. Baranowski, Warsaw 1854 (together with the Latin text) (27).

Translated by M. Brożek, with commentary by A. Birkenmajer Warsaw 1953; (together with the Latin text of Book I) (142).

German Translation

Translated and with commentary by L. C. Menzzer, *Über die Kreibeuegungen der Weltkörper*, Thorn 1879 (16); Reprint: Leipzig 1939 (121).

French Translation

Translated and with commentary by A. Koyré, *Des Revolutionibus des orbes célestes*, Paris 1934 (Translation of Book I to the end of chapter 11) (105). 2nd edition – Paris 1970.

English Translation

Translated by C. G. Wallis, *On the Revolutions of the Heavenly Spheres*, Chicago 1952 (138 a).
Translated by J. F. Dobson and S. Brodetsky, *De revolutionibus*, Preface and Book I. Occasional Notes. Royal Astronomical Society. no. 10: 1947 (134).

Russian Translation

Translated and with commentary by J. N. Veselovsky, *O vraštšenyah nebesnyh sfer*, Moskva 1964.

Spanish Translation

Translated by M. Taguena Lacorte and C. Moreno Canada, *Revoluciones de las orbitas celestes*, México 1969.

B. DE LATERIBUS ET ANGULI TRIANGULORUM...

(Vittembergae 1542, J. Luft. (Trigonometry) (2).

C. OTHER ASTRONOMICAL WRITINGS BY COPERNICUS

Two short dissertations preserved only in 16th century copies:

(1) *De hypothesibus motuum coelestium a se constitutis commentariolus.*

Latin Editions

in: *Inedita Copernicana* (ed. by M. Curtze), Toruni 1878. *Mitt-en Cop. Ver.* I. (55);
De hypothesibus... (ed. by A. Lindbagen), Stockholm 1881. *Svenska vet Akad. Handlingar* **6**, No. 12 (58)
in: Prowe, L., *Nicolaus Coppernicus* **2**, Berlin 1884 (63);
in: *Erster Entwurf seiner Weitsystems*... (ed. by F. Rossmann), München 1948 (135). Latin text with corresponding German translation.

English Translation

– in: Rosen, E., *Three Copernican Treatises*, New York 1959.

(2) *De octava sphaera contra Wernerum. Rev. Dom. Bernardo Wapowski* (known as *The Letter to Wapowski*)

Latin Editions

in: *De revolutionibus*... Warsaw 1854 (27);
in: Prowe, L., *Nicolaus Coppernicus* **2**, Berlin 1884 (63).

English Translation

in: Rosen, E., *Three Copernican Treatises*, New York 1959.

D. ECONOMICS

(1) Monetae cudendae ratio

Latin Editions

Dissertatio de optima monetae cudendae ratione, anno MDXXVI scripta nunc primum ex eius autographo typis vulgata (Ed. by Bentkowski), Warsaw 1816. *Pamiętnik Warszawski* (The Warsaw Memoirs) vol. V., and off-print (10) (Title and text also in Polish);
in: Prowe, L., *Monumenta Copernicana*, Berlin 1873 (46);
in: *De revolutionibus*... Warsaw 1854 (27);
in: *Spicilegium Copernicanum* (ed. F. Hipler), Braunsberg 1873 (48);
in: *Rozprawy o monecie i inne pisma ekonomiczne* (*Treatises on Money and other Economic Writings*), (ed. by J. Dmochowski), Warsaw 1924 (96); (Title and text in Latin and Polish).

Polish Translations

Dissertatio de optima monetae... Warsaw 1816 (10);
Rozprawy o monecie... (Treatises on Money), Warsaw 1924 (96).

French Translation

translated by W. Żółtowski; in: *Écrits notables sur la monnaie*, Paris 1934, vol. 1 (106).

English Translation (large excerpts)

in: Taylor, J., *Copernicus on the Evils of Inflation*, London 1955, *Journ. Hist. Ideas* 5.

(2) Early versions of the treatise on money from 1517, 1519, 1522 were published in:
Acta Tomiciana 5, 167–169 pp (1517 version);
Birkenmajer, A., *Stromata Copernicana*, Cracow 1924 (492);
Rozprawy o monecie (*Treatises on Money*) Warsaw 1924 (with a Polish translation) (96); *Spicilegium Copernicanum*... Braunsberg 1873 (48) (the 1519 and 1522 versions).

E. OTHER COPERNICAN TEXTS (PRINCIPALLY LETTERS)

The original manuscripts are found in the libraries and archives of Cracow, Olsztyn, Göttingen and Stockholm.

Latin Editions

in: *De revolutionibus*... Warsaw 1854 (27)
in: Prowe, L., *Monumenta Copernicana*, Berlin 1873 (46);
in: *Spicilegium Copernicanum* (ed. by F. Hipler), Braunsberg 1873 (48);
in: Biskup, M.: *Nowe materiały do działalności publicznej Mikołaja Kopernika z lat 1512–1537*, Studia i Materiały z Dziejów Nauki Polskiej Seria C., zesz. 15, 1971 (New Materials from the Public Activity of Nicolaus Copernicus in 1512–1537, Studies and Material on the History of Polish Learning Series C, no. 15, 1971).

F. SIMOCATTA THEOPHYLACTUS SCHOLASTICUS

Epistolae morales, rurales et amatoriae (translated from the Greek into Latin by Nicolaus Copernicus).
Latin Edition – Cracoviae 1509, J. Haller (1);
Facsimile of translation by Copernicus, Greek and Latin texts (ed. by R. Gansiniec), Polish translation – J. Parandowski – Warsaw 1953 (139).

2. Principal Works on Copernicus

(1) Prowe, L.: *Nicolaus Coppernicus*, Berlin 1883–1884, vol. 1–2 (429).
(2) Birkenmajer, L. A.: *Mikołaj Kopernik*, Cracow 1900, part 1 (44).
(3) Birkenmajer, L. A.: *Stromata Copernicana*, Cracow 1924 (492).
(4) Kuhn, T. S.: *The Copernican Revolution*, Cambridge 1957.
(5) Ravetz, J. R.: *Astronomy and Cosmology in the Achievement of Nicolaus Copernicus*, Wrocław-Warsaw 1965.
(6) Dobrzycki, J. and, Hajdukiewicz, L.: 'Mikołaj Kopernik', in *Polski Słownik Biograficzny* (The Polish Biographical Dictionary) **XIV**, No. 1 Wrocław-Warsaw-Cracow 1968.
(7) Rosen, E.: 'Nicolaus Copernicus', in *Dictionary of Scientific Biography* **III**, New York 1971
(8) Birkenmajer, A.: *Études d'histoire des sciences en Pologne*, Wrocław 1972. *Studia Copernicana* **IV** (The publication also contains treatises on Copernicus).
(9) Biskup, M.: *Regesta Copernicana*, Warsaw 1972.

NAME AND GEOGRAPHICAL INDEX

BARBARA BIEŃKOWSKA

ABOUT THE AUTHORS

STANISŁAW HERBST (b. 1907), professor at the University of Warsaw, head of the Faculty of Modern Polish History, chairman of the Polish Historical Society. Historian, respected authority on the history of culture and the history of cities.

JERZY DOBRZYCKI (b. 1927), associate professor at the Faculty of the History of Science and Technology of the Polish Academy of Sciences, deputy director of the Copernican Research Center. Astronomer, historian of astronomy, leading authority on Copernican works.

ALEKSANDER BIRKENMAJER (1890–1967), Poland's most outstanding Copernicologist, historian of the exact sciences and bibliologist.

WŁODZIMIERZ ZONN (b. 1905), professor at the University of Warsaw, director of the Institute of Astronomy of the University of Warsaw, well known popularizer of astronomical problems, astronomer.

LEOPOLD INFELD (1898–1968), well known theoretical physicist, worked with A. Einstein for many years.

WALDEMAR VOISÉ (b. 1920), professor, deputy director of the Faculty of the History of Science and Technology of the Polish Academy of Sciences, well known methodologist and historian of culture.

BOGDAN SUCHODOLSKI (b. 1903), professor, head of the Faculty of the History of Science and Technology of the Polish Academy of Sciences, member of the Polish Academy of Sciences, remarkable educator and historian of culture.

EDWARD LIPIŃSKI (b. 1888), professor emeritus of the University of Warsaw, member of the Polish Academy of Sciences, honorary chairman of the Polish Economic Society, leading economist and historian of economic thought.

BARBARA BIEŃKOWSKA (b. 1934), associate professor of he Faculty of Library Science at the University of Warsaw, historian of culture.

GPSR Compliance

The European Union's (EU) General Product Safety Regulation (GPSR) is a set of rules that requires consumer products to be safe and our obligations to ensure this.

If you have any concerns about our products, you can contact us on ProductSafety@springernature.com

In case Publisher is established outside the EU, the EU authorized representative is:

Springer Nature Customer Service Center GmbH
Europaplatz 3
69115 Heidelberg, Germany

Batch number: 10614746

Printed by Printforce, the Netherlands